NASA SP-388

ANALYTICAL APPLICATIONS OF BIOLUMINESCENCE AND CHEMILUMINESCENCE

Presented at the Second Annual Meeting
of the American Society for Photobiology
Vancouver, British Columbia, 1974

Compiled and edited by

Emmett W. Chappelle and Grace Lee Picciolo
Goddard Space Flight Center

Prepared by NASA Goddard Space Flight Center

Scientific and Technical Information Office   1975
NATIONAL AERONAUTICS AND SPACE ADMINISTRATION
Washington, D.C.

This document makes use of international metric units according to the Systeme International d'Unites (SI). In certain cases, utility requires the retention of other systems of units in addition to the SI units. The conventional units stated in parentheses following the computed SI equivalents are the basis of the measurements and calculations reported.

DEDICATION

This document is dedicated to the late Valerie N. Bush, one of the participants at the 2nd Annual Meeting of the American Society for Photobiology, who died on November 30, 1974. Mrs. Bush, an Instructor in Biology at Delaware State College, was for several years an extremely active researcher in bioluminescence applications. She contributed to the development and synthesis of the GSFC program for the application of luminescence to public sector needs. Her impact was felt especially in the area of fluctuations in bacterial ATP as a function of growth conditions and environment.

Valerie's enthusiasm, dedication, thoroughness, cheerfulness, and courage, even in the face of disability and death, are a lasting example to all who knew and loved her.

MESSAGE FROM THE PRESIDENT

On behalf of the American Society for Photobiology, I wish to thank all those individuals responsible for arranging the timely session devoted to applications of chemiluminescence and bioluminescence. This session is in keeping with the overall purpose of the Society which is to encourage and promote meetings and publications in all areas of photobiology.

I am both impressed and surprised at the increasing number of applications of chemiluminescence and bioluminescence, areas which a few years ago were solely of theoretical interest.

John Spikes
President
American Society for Photobiology

PREFACE

Bioluminescence and chemiluminescence are phenomena, which like so many natural phenomena, have passed through various stages of interest. They were first observed and admired and in some instances became the basis of certain superstitions. In the case of bioluminescence, the second stage was one of taxonomy or classification of the responsible organisms. The next stage for both bioluminescence and chemiluminescence was one of defining the reaction mechanisms—a task likely to continue for a very long time.

There are, however, some chemiluminescent and bioluminescent reactions where the reactants, products, and kinetics have been established with a high degree of certainty. Among these are the luminol or acyl hydrazide reaction, the bioluminescence of the firefly, and the bioluminescent reaction occurring in *Aequorea*.

This NASA Special Publication is a collection of the papers presented in a session on Bioluminescence Applications which took place at the 2nd Annual Meeting of the American Society for Photobiology, University of British Columbia, Vancouver, B.C., on July 25, 1974. They represent what is perhaps the final stage of the investigation of a natural phenomenon—its application for some useful purpose. They describe the exploitation of firefly bioluminescence for the measurement of adenosine triphosphate (ATP) in a variety of biological systems, the use of the bioluminescence of *Aequorea* in the sensitive assay of calcium ions, the ability of porphyrins to catalyze the chemiluminescent reaction of luminol (thus providing a detection method for bacteria), and finally the design of an instrument for low level light detection.

The high level of sensitivity and specificity inherent in the use of bioluminescent reactions for analytical purposes has led to a rapid proliferation of applications for these reactions as evidenced by the papers presented here.

This session, devoted to the applications of bioluminescence, is the second such conference taking place on this continent. The first was held during the International Symposium on Chemiluminescence and Bioluminescence at Athens, Georgia, in October 1972. The proceedings of that symposium were published by Plenum Press, N.Y., 1973, under the title of *Chemiluminescence and Bioluminescence*, M.J. Cormier, D.M. Hercules, and J. Lee, editors.

A more recent symposium on analytical applications of bioluminescence was held in San Diego, California, on March 4 through March 6, 1975, under the sponsorship of SAI Technology Co. (formerly the JRB Co.). The symposium, entitled ATP Methodology, covered applications of the firefly luciferase reaction including clinical bacterial detection, ocean biomass and marine sediments, vaccine viability, ATP content of sickle cells, seed viability, adenylate energy charge ratios, activated sludge, monitoring of biocide effectiveness, and an evaluation of commercially available photometers for bioluminescent measurements. The use of the bioluminescent reaction of photobacterium for the measurement of flavins (FMN, FAD, and riboflavin) and microorganisms was also described at this meeting.

Among the various procedures described was the firefly luciferase assay for ATP, developed at NASA/Goddard Space Flight Center. A manual covering the detailed procedures, "A Laboratory Procedures Manual for the Firefly Luciferase Assay for Adenosine Triphosphate" (NASA TM X-70926) was written by Emmett W. Chappelle and Grace Lee Picciolo in 1975.

The following papers in this publication reflect the continuing work and variety of applications for bioluminescence and chemiluminescence.

Emmett W. Chappelle, Session Chairman
Grace Lee Picciolo
Code 726
Goddard Space Flight Center
Greenbelt, Maryland 20771
July 1975

ACKNOWLEDGMENT

We wish to express our gratitude to the participants for their contributions
to the meeting's success and for their cooperation in providing us with
manuscripts of their presentations. We would like to thank Dr. John Lee
of the University of Georgia for his assistance in organizing the sessions,
and to the American Society for Photobiology, who sponsored the meeting,
goes our appreciation for their permission and cooperation in publishing
these papers.

EWC
GLP

CONTENTS

CONTENTS (continued)

PROBLEM AREAS IN THE USE OF THE FIREFLY LUCIFERASE ASSAY FOR BACTERIAL DETECTION

Grace Lee Picciolo and Emmett W. Chappelle
Goddard Space Flight Center
Greenbelt, Maryland

Elizabeth A. Knust and S. A. Tuttle
New England Medical Center
Boston, Massachusetts

C. A. Curtis
Hahnemann Medical College
Philadelphia, Pennsylvania

INTRODUCTION

Every child has enjoyed the spectacle of flashing yellow-green lights from fireflies on warm summer nights; now the understanding of the chemical nature of this light production, known as bioluminescence, has enabled scientists to apply this phenomenon of cold light to many fields and this includes determination of the presence and quantity of bacteria. Building upon E. Newton Harvey's (1965) extensive and enthusiastic investigations of numerous types of luminescences displayed by organisms in nearly every phylum in the living world, McElroy and his associates (1969) and subsequent workers extracted the light producing substances from fireflies and described the reaction mechanism and kinetics as follows (Plant et al., 1968):

$$E + LH_2 + ATP \xrightarrow{Mg^{++}} E \cdot LH_2 \cdot AMP + PP$$

$$E \cdot LH_2 \cdot AMP + O_2 \longrightarrow E + AMP + CO_2 + h\nu + T$$

where:

E = firefly luciferase enzyme

LH_2 = reduced luciferin

ATP = adenosine triphosphate

AMP = adenosine monophosphate

PP = pyrophosphate

T = thiazolinone

hν = light (550 nm)

The amount of light produced is proportional to the reactants when each is limiting. A light measurement capability could be used as an assay method for any of the reactants. Since ATP is a metabolite significant in all energy exchanges within living cells, its assay has implications for many parameters that describe the biota. ATP is known to be present in all forms of life, so its measure could establish the presence of living things. By purifying the firefly luciferase extract and adding all necessary chemicals but ATP in excess, an assay for ATP can be performed by measuring the amount of light produced when a sample containing soluble ATP is added to the luciferase reaction mixture. Thus the amount of light is proportional to the amount of ATP added (Strehler, 1965). When an unknown sample containing living organisms is processed to remove all exogeneous ATP, and the ATP within the organisms is released, a measure of the resultant light indicates the amount of organism ATP. If it is further demonstrated that within groups of organisms, their ATP content is relatively constant, then luciferase-ATP-initiated-light could be used to estimate the numbers of these organisms. This has been shown for bacteria, and procedures have been established to use the luciferase method to quantitate bacteria (Chappelle and Levin, 1968; D'Eustachio and Levin, 1967).

NASA/Goddard Space Flight Center personnel developed procedures for the determination of extraterrestrial life on other planets several years ago and subsequently evolved a program for applying these technological developments to public sector needs, particularly in the area of Health Care Delivery. One area, that of infection detection in urine specimens, is described in this paper.

The paper by Dr. Vellend describes an extension of this work, which was suggested by Dr. David Rutstein of the Harvard Medical School, for the determination of antimicrobial susceptibilities.

INSTRUMENTATION

Many types of photometric instruments exist for quantitating light. Those suitable for discriminating changes in low levels of light as is emitted by the small amounts of ATP present in bacteria, picomoles (10^{-9} moles) of ATP, employ a photomultiplier (PM) tube as a detector and a d.c. amplifier or equivalent. Several companies manufacture light measuring instruments including Schoeffel, Inc., Gamma Scientific, Inc., and Photovolt, Inc.; however, only a few have provisions for measuring light from solutions, such as Scintillation Counters, Packard, Inc., and Technicon, Inc., while those of American Instrument Co. (Aminco), DuPont de Nemours, and JRB, Inc. (SAI, Inc.), provide a means to inject the sample into the luciferase while both are in a light-tight configuration in place before the photocathode surface. Some investigators have assembled their own instruments from off-the-shelf amplifiers and photomultipliers with the addition of a reaction chamber with capability for injection of liquid (Chappelle and Levin, 1968).* A sensitive instrument has been produced that uses pulse counting. In this instrument, "Diogenes," the PM tube can be cooled to improve the signal-to-noise ratio, an automatic injection system is provided, and there is selection of analog or digital readout (Chappelle and Levin, 1966). With this instrument, 10^{-9} μg of ATP produces a signal above the noise but it is not linear. Because the light from unknown samples may differ by several decades, an automatic electronic amplification switching capability or digital readout in several digits is very useful and eliminates repeating the assay and thereby using more of the luciferase reaction mixture. Instruments that provide this capability are the DuPont Biometer, scintillation counters, the JRB ATP Photometer, and the Aminco Chem-Glow with integrator; while Hewlett Packard makes a picoammeter with automatic ranging that can be coupled with a photomultiplier tube with its power supply.

The light production upon injection of a sample containing ATP into the luciferase reaction mixture rises to a maximum intensity and then decays exponentially. Both the maximum intensity (determined by measuring peak height) and the total light output are proportional to the amount of ATP added. Therefore, several types of data display are suitable. When used in conjunction with a capacitance circuit, the peak height can be sampled and a digital output produced and recorded. A rate function can be produced on a strip chart recorder and a peak height measured. Analog circuitry can be introduced, digitized, and printed to show the total light output. Both the scintillation counter and one version of the JRB instrument use a measurement made during the decay period, since there is no provision for injecting sample into luciferase while positioned in front of a PM tube. However, the JRB provides an attachment for making this modification and also for reading the peak light. As an outgrowth of space research, originally intended for the detection of life on Mars, a prototype instrument was developed to perform chemical processing of a sample with subsequent injection of the luciferase automatically (Picciolo, 1971).

APPLICATIONS

Numerous areas of application are appropriate for using the luciferase assay for ATP of itself as well as to measure organism levels. Since it is specific for ATP and can be performed in the presence of other purine compounds, it does not require isolation of the ATP, so it could be measured in cell extracts

---

* See Wampler, J.E. paper in this document.

and body fluids. The ATP level is an indicator of metabolic changes in living organisms and therefore can be a monitor of cell integrity, genetic variants, substrate utilization (Klofat et al., 1969), attack by a virus (Chappelle and Levin, 1968; Levin et al., 1964), and cell growth. With further research, these types of measurements could have applications in cancer research (Vlodavsky et al., 1973; Levin et al., 1964) and immunology as well as organ and tissue viability and transplant rejection.

When used as a measure of bacterial levels, the ATP assay can be used for pollution monitoring in air and water supplies, such as drinking water, sewage treatment effluent, river and stream pollution levels, and industrial water supplies and effluents. Closed environment monitoring could be performed for spacecraft, space stations, clean rooms, operating rooms, and so forth. Microbial levels could be measured in dried foods, cereals, spices, milk, beer, wine, liquors, canned and bottled foods, pharmaceuticals, cosmetics, ointments, creams, paints, gasolines, and oils. Agricultural uses include determination of fertility levels, spore viability, and sterility of soils, plants, and animals. Oceanographic monitoring can be performed to determine biomass and effect of pollutants. Medical applications include infection levels in blood, urine, cerebrospinal fluid, wound excretion, joint fluid, and lung and pleural fluid. Evaluation of antibiotic effects on bacteria including synergism as well as antibiotic levels in body fluids could be done using the luciferase assay.

Each of these applications represents areas where research is needed into the problems involved in making the luciferase assay applicable. These include sampling, sensitivity, and background. The advantages in most cases would be speed and specificity.

BASIC CHARACTERISTICS OF THE LUCIFERASE ASSAY

Since the efficiency of light production in the firefly has been shown to be near unity (McElroy and Glass, 1961), it is expected that the in vitro reaction could be made fairly efficient also, thus providing sufficient light signal when measuring small amounts of the other reactants. When all of the chemicals necessary to the reaction are present in excess and when the limiting chemical has been removed from the luciferase reaction mixture, the addition of that chemical stimulates the production of light that is quantitatively related to the concentration of that chemical. When that chemical is ATP, it has been demonstrated that both the peak light output, which occurs within one-half second after proper mixing, and the total light output are proportional to the amount of added ATP (Chappelle and Levin, 1968; St. John, 1970). Within practical limits of reagent concentration and instrument amplification, ranges of ATP can be measured from 5×10^{-2} to $5 \times 10^{-7} \mu M/ml$. With careful manipulation, linearity and reproducibility will range between 5 and 25 percent coefficient of variation, where measuring standard ATP in the absence of inhibitors can be done very accurately, while measurements of bacteria from unknown specimens produce the most fluctuation.

4

Under ordinary growth conditions, changes in ATP/bacterium have been shown to vary by about a factor of 10 (figure 1). However, for each type of specimen, this variation should be checked if it is considered that the environmental conditions could affect the ATP/bacterium.* Various conditions have been shown to lower the ATP/bacterium; removing O_2 from the medium gives a rapid, dramatic decrease in ATP levels (Klofat et al., 1969).

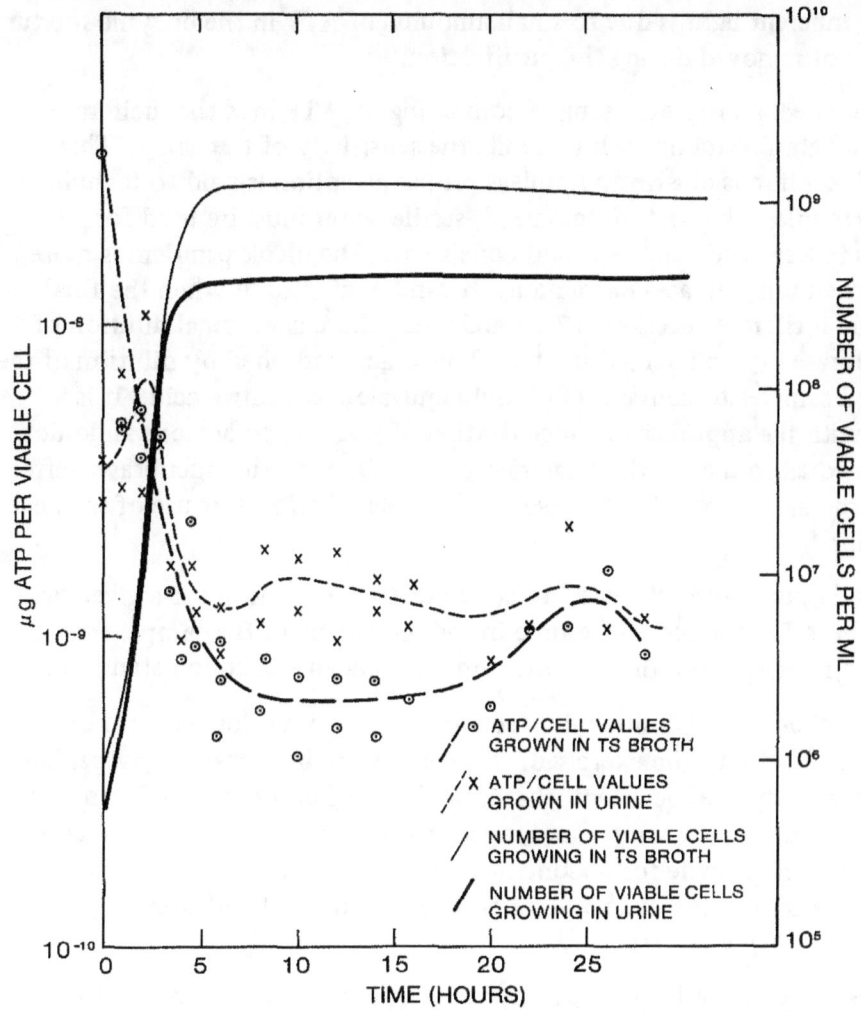

Figure 1. Effect of the growth medium and the *length of time* grown in this medium on ATP per viable *E. coli* values.

With proper handling, the luciferase reaction mixture is reliable and reproducible. When purification of the luciferase is carried out consistently, the activity can be standardized from batch to batch. The activity decays

---

* See Holm-Hansen et al. paper in this document.

5

only slightly at room temperature for up to two hours. Care must be exercised that the sample and luciferase are at the same temperature at measurement.

The luciferase reaction mixture emits light without the addition of any ATP. This is called inherent light. Preincubation will reduce the level of inherent light as well as action of an ATPase, but since the activity also decreases, a compromise time must be selected to achieve an optimum level. It is suggested that the inherent light is due to small amounts of ATP in the enzyme mixture that are not removed during the purification.

When we inject a processed sample containing no ATP into the luciferase, there is a light production which limits the sensitivity of the assay. This "blank" reaction is elusive and, unless proper attention is paid to technique, can be variable. The best of deionized, sterile water must be used for all reagents to keep the blank low and consistent. The blank problem is more severe when using an acid extractant. It can be minimized when the final pH of the luciferase reaction is 7.75 and when there is minimal dilution of the luciferase by the injected sample. This is accomplished by dilution of the extracted samples to contain 0.005 milliequivalents of nitric acid. It is diluted with the appropriate concentration of Na_2SO_4 to achieve an ionic strength equal to that of the luciferase buffer. The starting luciferase buffer pH and molarity is selected to ensure a final pH of 7.75 after use of the acid extractant.*

Chemically purified ATP is used to standardize the reaction for a given light level. The ATP is added at the time in the processing of the sample that it would be released from the bacteria, representing an extraction standard.

Since the chemical environment affects the light production, control of ionic strength and pH must be exercised. This effect can be measured by performing a recovery-type experiment, that is, adding higher amounts of standard ATP after assaying the sample and repeating the assay. Mathematical adjustment can then be made for uncontrollable variations in the chemical environment. This is the same method that is called an internal addition standard (St. John, 1970).

The sensitivity of the luciferase assay is determined by the activity of the luciferase under the specific chemical assay conditions and is limited by the blank response. Increase in ionic concentration decreases the activity and increases the blank response. Specific ions stimulate or inhibit activity and blank response correspondingly.

The response of commercially prepared ATP of various concentrations minus the blank response is plotted as a function of ATP concentration in figure 2. Reproducibility, sensitivity, and linearity of the luciferase assay is given in table 1. The disodium salt of ATP is diluted in distilled water, and the purified luciferase is supplied by DuPont, Inc., and is reconstituted in 0.01 M

* Chappelle, E. W. and G. L. Picciolo, C. A. Curtis, E. A. Knust, D. A. Nibley, R. B. Vance, "Laboratory Procedures Manual for the Firefly Luciferase Assay for Adenosine Triphosphate (ATP)," NASA-GSFC TM X-70926, 1975.

6

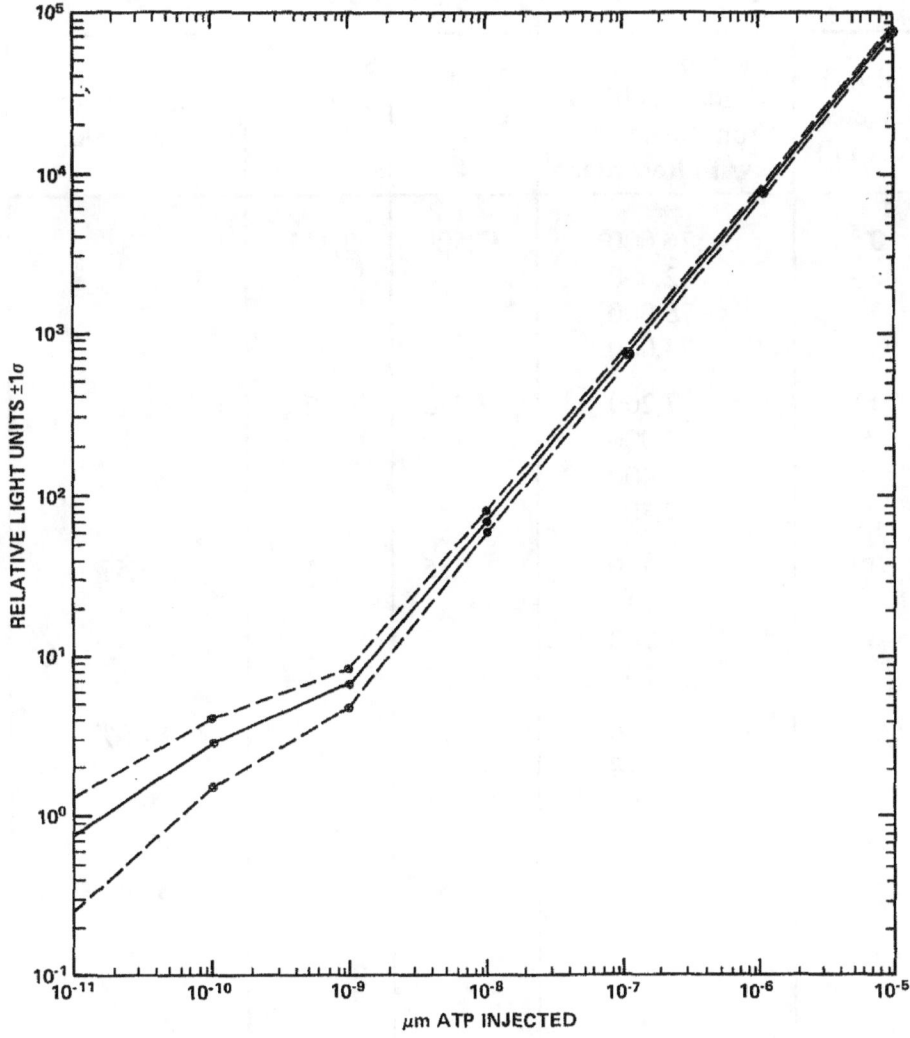

Figure 2. ATP concentration curve showing relative light units ±1σ versus μM ATP injected (0.1 ml) using the Aminco Chem-Glow Photometer with recorder. The ATP is diluted in water and the DuPont luciferase is reconstituted in 0.01 M TRIS, 0.01 M MgSO₄, pH 8.0, using 0.1 ml/cuvette.

TRIS buffer and 0.01 M MgSO$_4$ at pH = 8.0. The conditions represent minimal inhibition and therefore maximum activity.

However, when chemical processing must be performed on the sample, reagents are added to the sample which result in inhibition and therefore decrease in acitivity, which may result in loss of sensitivity. Table 2 shows the linearity and sensitivity of the response to standard ATP when diluted in saline, broth, and urine. This may be compared with the uninhibited condition (table 1). The conditions used here would be used when measuring bacterial ATP in the presence of mammalian cells and other nonbacterial

Table 1

Data for Figure 2*

μM ATP Injected (0.1 ml)	Relative Light Units Less Blank From Chem-Glow with Recorder	MEAN $\overline{X}$	Standard Deviation σ	Coefficient of Variation
10^{-5}	76,000 82,000 78,000 74,000	77,500	3,415	4%
10^{-6}	7,200 7,000 7,400 7,500	7,275	222	3%
10^{-7}	780 740 740 760	755	19	2.5%
10^{-8}	78 78 68 68	73	5.8	7.9%
10^{-9}	5 9 6 7	6.7	1.7	25%
10^{-10}	3 3 1 4	2.7	1.2	46%
10^{-11}	1 0 1 1	0.7	0.5	66%

* The dotted line dividing the data indicates the cutoff point below which the results were not linear and were not used in linear regression analysis.

Linear regression analysis showed: using logarithms of the numbers when n = 20, the slope interval is 0.97 < 0.98 < 0.99; the intercept is close to 0; the F ratio = 1.7×10^4 when table $F_{0.95}$ = 19; and correlation coefficient (r) is 0.999.

Table 2
Linearity and Sensitivity Response to Standard ATP

A. Results from the Centrifuged Procedure using Commercial ATP Diluted in Saline, Trypticase Soy Broth, or Urine				
μM ATP/ml Original Sample	μM ATP Injected	Light Units Less Blank		
		Saline	Broth	Urine
4.4×10^{-4}	1.1×10^{-3}	Saturated	Saturated	Saturated
4.4×10^{-5}	1.1×10^{-4}	62,000	58,000	54,000
4.4×10^{-6}	1.1×10^{-5}	6,000	6,000	6,500
4.4×10^{-7}	1.1×10^{-6}	600	550	600
4.4×10^{-8}	1.1×10^{-7}	58	70	54
4.4×10^{-9}	1.1×10^{-8}	28	15	7
4.4×10^{-10}	1.1×10^{-9}	2	4	0
B. Results from the Nonconcentrated Procedure using Commercial ATP Diluted in Saline, Trypticase Soy Broth, or Urine				
1×10^{-2}	1×10^{-4}	45,000	55,000	52,000
1×10^{-3}	1×10^{-5}	6,500	6,500	5,400
1×10^{-4}	1×10^{-6}	700	600	530
1×10^{-5}	1×10^{-7}	61	45	44
1×10^{-6}	1×10^{-8}	22	5	9
1×10^{-7}	1×10^{-9}	2	0	4
1×10^{-8}	1×10^{-10}	4	0	9
1×10^{-9}	1×10^{-11}	0	0	3

sources of ATP. Results are given when reagents are allowed to remain in the sample (nonconcentrated procedure), and when they are mostly removed by the centrifugation procedure. On the basis of actual ATP injected, the light response for the two methods is approximately the same. It thus follows that in the centrifugation procedure, where one injects the equivalent of 2.5 ml of original sample instead of 0.01 ml as in the nonconcentrated procedure, the overall sensitivity should increase by about 250 times. This is shown in the first column of table 2, micromoles of ATP per milliliter of the original sample. The upper level of ATP which can be measured is limited by the saturation of the photocathode of the photomultiplier or by the concentration of other reagents, such as luciferase and luciferin. The lower level is limited by the blank level which must be subtracted from each reading. One area of work is to develop a means of measuring a blank for each unknown sample or eliminating the blank. Since the nature of the sample can produce variation in blank levels, this is a very critical and difficult problem.

The presence of each of the assay reagents introduces inhibition of the luciferase activity in varying amounts. For a constant amount of ATP, the percent of inhibition for each of the complete procedures when compared with a procedure with no inhibiting reagents included for ATP in H_2O, luciferase in 0.01 M TRIS, and 0.01 M $MgSO_4$ is:

Nonconcentrated with malate buffer
    ATP in saline   42%
    ATP in urine   33%
Centrifuged with malate buffer
    ATP in saline   20%
    ATP in urine   30%

When it is expected that fluctuations will be introduced by the unknown samples, each sample must be evaluated for its inhibition. This is done by assaying the unknown and then adding a negligible volume of standard ATP at a concentration far in excess of the sample ATP and repeating the assay. By this recovery method, an exact correction can be made for all inhibitors known and unknown in each sample. Coefficient of variation on the inhibition of individual urine specimens is about 32 percent. The sample reading is subtracted from the recovery reading and then the ratio of sample reading to net recovery reading is equivalent to the ratio between the amount of ATP in the sample and the amount of ATP added to the recovery.

In order to quantitate bacteria in a sample by measuring the ATP in that sample, the ATP from nonbacterial sources must not interfere. The effective ratio of bacterial ATP to nonbacterial ATP is small and is a limit on the resolution of the analysis. Biological fluids may contain soluble ATP and also ATP contained within nonbacterial cells such as mammalian cells in a urine specimen. A description follows of the procedure we have developed to measure bacterial ATP in biological fluids such as urine specimens.

FUNCTION OF REAGENTS USED IN THE PROCEDURE

Samples containing tissue cells can be processed by the addition of a nonionic detergent such as Triton X-100, which releases tissue cell ATP without affecting bacterial ATP at the concentration used. This was verified using 0.17 percent Triton X-100 on the group of urinary bacteria listed below. This was done with a higher Triton X-100 concentration than is used in the procedures. The only organism that gave a significant loss of ATP was β

Streptococcus. A concentration of 0.1 percent Triton X-100 will release the ATP from the number of blood cells contained in 10 percent blood. Other types of tissue cells could be present in a urine sample such as epithelial and kidney cells. We subjected 1×10^6 monkey kidney cells per ml to the Triton X treatment and removed the released ATP by apyrase. Subsequent treatment with perchloric acid recovered up to 9 percent of the kidney cell ATP.* This was done without the pH drop that we found necessary to remove bound ATP from blood cells; therefore, we will repeat this experiment incorporating the malic acid procedure which is the pH drop.

Effect of 0.17 percent Triton X-100 on Bacterial ATP from 10 Species of Bacteria

Organism	Loss of ATP
Proteus mirabilis	0 %
Proteus morganii	0 %
Serratia marcescens	0 %
Streptococcus faecalis	0 %
Staphylococcus aureus	0 %
Herellea species	8.8%
Pseudomonas aeruginosa	0 %
Klebsiella pneumoniae	12 %
Escherichia coli	0 %
β Streptococcus	90 %

Any ATP that is present in solution can be removed by the addition of an ATPase, such as potato apyrase, which hydrolyzes the ATP to a form that is inactive in the luciferase reaction. Potato apyrase requires Ca^{++} as a cofactor at a reacting concentration of 5×10^{-3} M when using up to 6.7-mg purified apyrase/ml of reacting solution. The activity of the Sigma purified apyrase used is given as 2.5 units per mg, where a unit is that amount that will liberate one micromole of inorganic phosphorus per minute at pH 6.5 at 303 K (30°C) from ATP. The optimal pH is 6.5, but there is some activity as low as pH 3.0.

Residual traces of ATP remain in the sample at this point; these are bound to large molecules (such as proteins) and particulates. This ATP can be removed by lowering the pH to dissociate ATP from its binding sites, allowing the apyrase to work at this time. Since the pK_4 of ATP = 4.0, we drop the pH to the lowest value possible to still retain marginal apyrase activity, pH = 3.75.† Malate buffer is used to buffer the sample down to this pH.

---

* Bush, V. N., "The Examination of Urine: Samples for Pathogenic Microbes by the Luciferase Assay for ATP. I. The Effect of the Presence of Fungi, Fungal-like Bacteria and Kidney Cells in Urine Samples," NASA Grant Report No. N73 18096, 1973.

† Since the work reported in this paper, we have changed the malate buffer to a pH = 4.25 whereby there is no rupture of urinary pathogenic bacteria.

The sample now contains only ATP contained within the bacterial cell. This is released by an acid extractant.

Organic solvents and inorganic acids were compared for their efficiency on some of the urinary bacteria. Conditions were established for acetone and nitric acid that gave the highest recovery.* For ten of the urinary pathogens, the effect of increasing concentrations of HNO_3 is shown in table 3 as the relative amount of ATP extracted, and is shown as a function of the milli-equivalents HNO_3 (on the pellet of stationary phase organisms). We then chose 0.1 N HNO_3 as the extracting concentration of the acid for urinary pathogens as a compromise between extraction efficiency and inhibition.

Table 3

Relative Amount of ATP Extracted from Various Bacteria using
Various Concentrations of Nitric Acid on the Pellet

Organism	Normality of Nitric Acid			
	0.0625	0.100	0.150	0.200
Escherichia coli	1.6	1.7	2.0	2.0
Klebsiella pneumoniae	1.9	2.1	2.0	2.0
Staphylococcus aureus	1.4	1.8	1.9	1.8
Pseudomonas aeruginosa	0.98	1.9	2.7	1.9
Proteus mirabilis	1.8	2.0	2.0	2.1
Enterobacter cloacae	1.2	2.1	3.7	3.3
Streptococcus faecalis	0.38	0.88	1.1	1.0
Serratia marcescens	88.1	91.5	60.3	66.4
Proteus vulgaris	77.6	99.3	79.6	78.4
Staphylococcus epidermidis	41.6	41.0	45.0	38.6

The use of the acid extractant also holds the pH below 2.0 which inactivates the apyrase. The sample is diluted so that the amount of acid injected into luciferase is 0.005 meq in a final volume of sample luciferase mixture of

---

* See Knust, Elizabeth A., et al., paper in this document.

0.2 ml with a final TRIS buffer concentration of about 0.13 M. The final pH of the sample luciferase mixture must be at 7.75. In order for this to hold, the starting pH of 0.25 M TRIS must be 8.2. The optimal range of pH after each reagent addition is given in the following list. The nonconcentrated procedure is given in procedure 1.

Range of pH Tolerance

After Addition to Sample of:	Acceptable pH Range
Apyrase-Ca ± Triton X-100	5.0 to 7.8
Malate buffer	4.00 ± 0.25
Nitric acid	1.15 to 1.20
H_2O or Na_2SO_4	1.8 to 2.1
Luciferase-luciferin	7.75 ± 0.1

Procedure 1

Nonconcentrated Method for Luciferase Assay of Bacteria: Malate-Nitric Acid Procedure

0.5 ml sample: Urine, bacterial culture, or other.

Add 0.1 ml apy-TX-Ca: 10 mg apyrase/ml 0.03 M $CaCl_2$
   (0.6% TX-100 if contaminating mammalian cells are present).
   Wait 15 minutes, vortexing frequently.

Add 0.1 ml malate buffer: 0.5 M malic acid 0.005 M Sodium Arsenate, pH 3.75.
   Wait 15 minutes, vortexing frequently.

Add 0.1 ml 1.5 N HNO_3
   Vortex well, wait 5 minutes.

Add 4.2 ml 0.15 M Na_2SO_4

Assay: Inject 0.1 ml of above into 0.1 ml of luciferase (DuPont), reconstituted with 1.5 ml 0.25 M TRIS, 0.01 M $MgSO_4$ pH 8.2 per vial.

Recovery: 0.05 ml of ATP (10 μg/ml) or 0.5 ml of ATP (1.0 μg/ml), depending on desired accuracy of delivery, is added to a measured amount (0.3 ml) of the remainder of the treated sample.

CONCENTRATION METHODS

By concentrating the sample one not only increases sensitivity, but removes endogenous inhibitors as well.

Standard ATP under these conditions would be assayed by the following procedures.

The reproducibility, sensitivity, and linearity for standard ATP concentrations, when assayed after the addition of 0.1 N HNO_3 and diluted 1:1 with H_2O and 0.1 ml of this mixture is injected into 0.1 ml of luciferase-luciferin mixture in 0.25 M TRIS, 0.01 M $MgSO_4$, pH = 8.25, are shown in table 4.

Two methods have been optimized using centrifugation: The pure species short centrifugation procedure (procedure 2) is designed to be used in bacterial studies where there are not large amounts of nonbacterial cells. It uses only one centrifugation. It is intended to take approximately 1 hour, thereby allowing greater ease in the execution of timed experiments as well as maintaining the accuracy of the longer centrifugation procedure.

Procedure 2

Pure Species Short Centrifugation Procedure

5.0 ml sample

Add 1.0 ml Tx-Apy-Ca: 10 mg/ml purified apyrase in 0.6% Tx-100, 0.03 M $CaCl_2$.
Vortex well.

Centrifuge 15 minutes at 8000 rpm (10400 RCF × G) and 20°C.
Decant inverted on filter paper 5 minutes.

Add 0.2 ml 0.1 N HNO_3.
Vortex, wait 5 minutes.

Add 0.2 ml H_2O, sterile, deionized.
Vortex.

Assay: Inject 0.1 ml into 0.1 DuPont luciferase reconstituted with 1.5 ml 0.2 M TRIS, 0.01 $MgSO_4$, pH = 8.4.

The second, longer centrifugation procedure (procedure 3) is used when the sample contains other than bacterial cells which interfere with ATP measurement. Any sample of biological fluid can be used that does not contain more than 10^6 leucocytes or more buffering capacity than 10 percent blood or 100 percent urine or 100 percent trypicase soy broth. Any volume up to the capacity of the centrifuge tube used can be accommodated, with proper adjustments for concentration.

14

Table 4

Reproducibility, Sensitivity, and Linearity for Standard ATP Concentrations When Assayed*

Instrument:	Biometer	
Procedure:	Analogous to the centrifuge procedure omitting all reagents and steps except nitric acid and the final water†	

μg ATP Injected	CV% (Blank not subtracted)	Light Units less blank †
3.5×10^{-1}	10%	8.9×10^9
3.5×10^{-2}	17%	9.12×10^8
3.5×10^{-3}	7%	1.09×10^8
3.5×10^{-4}	7%	9.29×10^6
3.5×10^{-5}	13%	1.29×10^6
- - - - - - - - - ‡		- - - - - - - - - ‡
3.5×10^{-6}	------	1.01×10^6
3.5×10^{-7}	------	6.70×10^5
Blank	0.15%	4.21×10^6

Linear Regression Analysis:

Slope interval, $\beta = 1.06 > 1.02 > 0.98$ and intercept $= -1.06 \times 10^1$ when $P = 0.05$ and $N = 15$; coefficient of correlation $(r) = 9.98 \times 10^{-1}$

F ratio $= 2.78 \times 10^3$ when table $F_{0.95} = 2.69$

* See text.

† Water was used as a diluent here, giving a higher blank than is expected when the diluent is Na_2SO_4.

‡ The dotted lines through the data are the cutoff points below which the results are not linear and were not used in linear regression analysis.

Procedure 3 involves the use of 17- $\times$ 100-mm disposable plastic test tubes of 12-ml capacity. Centrifuge well adaptors were fabricated to fit a 250-ml-volume well drilled to hold five plastic tubes per well. This allows a total of 30 tubes per centrifugation.

The light measured is calibrated by adding known amounts of chemically pure ATP to a duplicate sample (internal addition standard). After subtracting the sample light value from the internal addition standard light value,

Procedure 3

Centrifugation Method for Luciferase Assay of Bacteria:
Malate-nitric Acid Procedure

10 ml sample: urine, bacterial culture, or other.

   0.2 ml 6% Triton X-100
   Vortex well.
   Centrifuge at 10,400 RCF $\times$ G, 293 K (20°C)

Add 1.0 ml Apy-Ca: 10 mg apyrase/ml 0.03 M $CaCl_2$.

   Vortex well.

Add 5.0 ml normal saline (0.9%).

   Mix well.
   Wait 15 minutes.

Add 1.0 ml malate buffer: 0.25 M malic acid 0.005 M Na.

   Mix well.
   Centrifuge at 10,400 RCF $\times$ G, 293 K (20°C).
   Discard supernatant and invert tube to drain on paper toweling for
      5 minutes.

Add 0.2 ml 0.1 N HNO_3.

   Vortex well.
   Wait 5 minutes.

Add 0.2 ml 0.15 M Na_2SO_4.

Assay: Inject 0.1 ml of above into 0.1 ml of luciferase (DuPont)
         reconstituted with 1.5 ml 0.25 M TRIS with 0.01 M $MgSO_4$
         pH 8.2 per vial.

Recovery: 0.05 ml of ATP 1.0 μg/ml is added to the remainder
            of the treated sample.

a direct proportion is made between the unknown sample and the known quantity of ATP added. The ATP value then can be converted to number of bacteria per milliliter by assuming an average ATP per bacterial cell or used directly as a measure of bacterial ATP levels.

Table 2 shows the results obtained when urine, saline, or broth are taken through the centrifugation procedure and dilutions of chemically synthesized ATP added and assayed. In order to summarize the reproducibility and sensitivity of the assay for detecting bacteria, we ran the following experiment where we determined *Escherichia coli* cell number by the luciferase centrifugation assay and the nonconcentrated assay, microscopic counting, and agar plate colony counts. *E. coli* cultures were grown to log phase in trypticase soy broth, centrifuged, and equal aliquots resuspended in normal saline, filtered, pooled urine and fresh trypticase soy broth. Microscopic and spread plate counts were run on all three aliquots. Serial dilutions were made on the three suspensions and the nonconcentrated and the long centrifugation procedures were run on the dilutions. The microscopic counts agreed closely with the plate counts indicating that all the bacteria were dividing and therefore were viable. Data in table 5 give the bacteria per milliliter of stock solution obtained by averaging the microscopic and plate counts, the coefficient of variation for each set of measurements, the ATP per milliliter stock solution, the number of bacteria per milliliter that gives a response of one unit above the blank value, and the average ATP per bacterium calculated from this experiment. The graphs in figure 3 show the relative light units by the luciferase assay versus bacteria per milliliter by microscopic and plate counts.

Table 5

Results from Long Centrifugation Procedure with *E. coli*
in Saline, Urine, and Broth (see text)

	Saline	Urine	Broth
Average of microscopic and colony counts-bacteria/ml	1.6×10^8	1.6×10^8	1.9×10^8
Coefficient of variation	18%	14%	19%
μM ATP/ml	3.0×10^{-5}	2.7×10^{-4}	5.4×10^{-4}
Coefficient of variation	25%	25%	19%
Sensitivity-bacteria/ml	3.6×10^3	4.1×10^2	2.3×10^2
μM ATP/bacterium	1.8×10^{-13}	1.6×10^{-12}	2.8×10^{-12}

The lowest number of cells which can be detected by the nonconcentrated procedure has been found to be between 5×10^5 and 1×10^6, depending on the species and media, as compared with 3×10^3 to 2×10^2 for the centrifugation procedure.

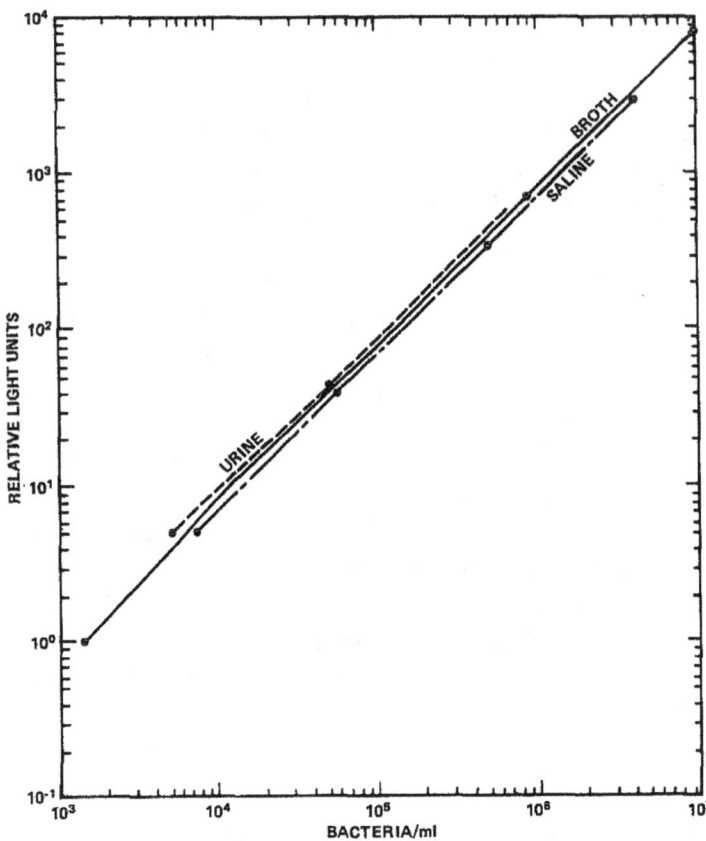

Figure 3. *E. coli* concentration curve showing relative light units versus bacteria per milliliter by microscopic and plate counts. The long centrifugation procedure was run on the bacteria diluted in urine (- - -), saline (— · —), and broth (—).

The noncentrifuged procedure would be more applicable because of its simplicity in situations where the cell number is high, that is, greater than 10^5, and where the media does not contain luciferase inhibitors, for example, pure cultures in defined growth media.

SENSITIVITY AND REPRODUCIBILITY OF THE ASSAY: SHORT CENTRIFUGATION

Using trypticase soy broth as samples, the short centrifugation procedure (procedure 2) was run and standard ATP added after use of nitric acid. The DuPont luciferase was reconstituted either with 0.15 M TRIS, 0.01 M $MgSO_4$ pH 8.5 or 0.2 M TRIS, 0.01 M $MgSO_4$, pH 8.4. Figure 4 shows the results when a 4-1/2-hour culture of *E. coli* is diluted in urine, saline, or broth and a short centrifugation procedure run, where large numbers of nonbacterial cells are not expected. Numbers of bacteria per milliliter, determined by Coulter Count of original sample, are plotted versus μg ATP per milliliter measured.

18

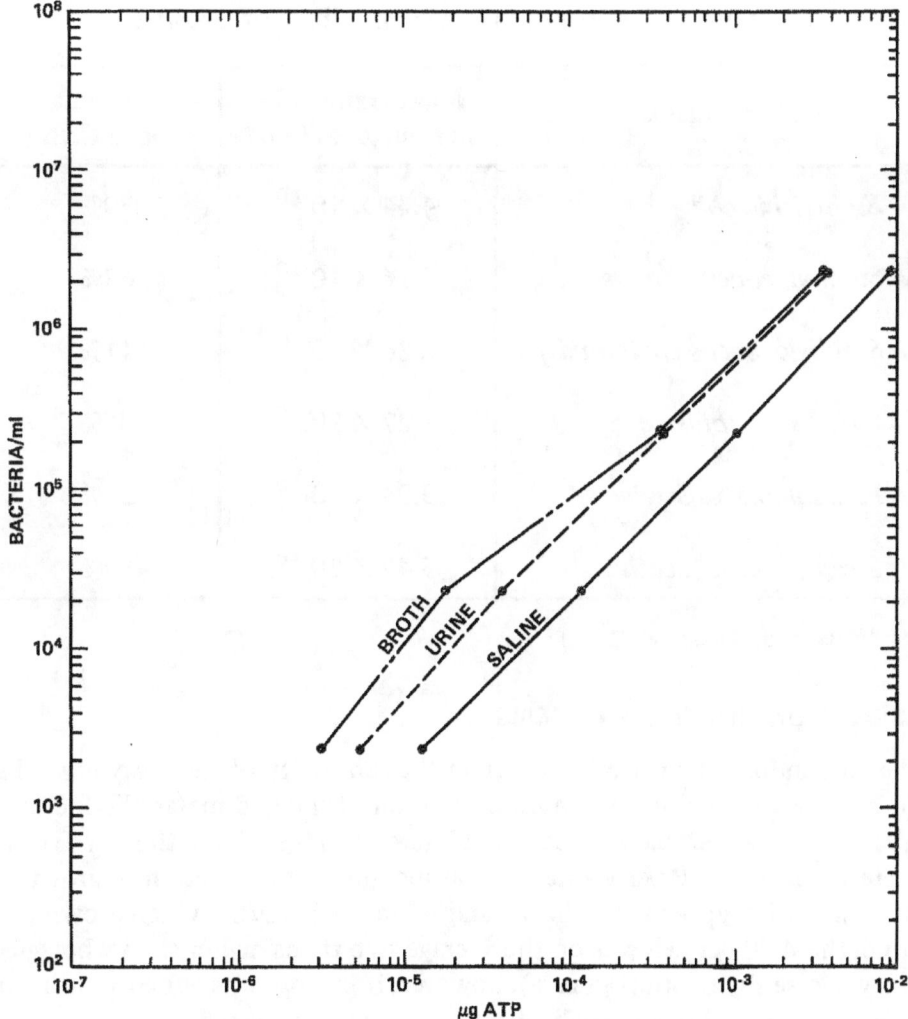

Figure 4. A 4-1/2-hour culture of *E. coli* was diluted in urine (- - -), saline (—), and broth (— - - —). The micrograms of ATP, measured using the short centrifugation procedure, are plotted versus the number of bacteria per milliliter as determined by Coulter Counter.

Table 6 shows the ATP per bacterial cell for some of the urinary pathogens. The bacteria were in stationary phase of growth and the centrifugation procedure was used to determine ATP levels. A microscopic or Coulter Count was done to determine total cell count, and drop plate colony counts were done using trypticase soy agar for viable counts. The percentage of dead cells was calculated as the difference.

Table 6

Urinary Pathogens in Stationary Phase of Growth

Microorganism	Micrograms ATP per Bacterial Cell*	Percent Dead Cells
Escherichia coli	4.44×10^{-10}	39%
Staphylococcus aureus	5.16×10^{-10}	63%
Staphylococcus epidermidis	2.26×10^{-10}	11%
Proteus mirabilis	1.27×10^{-10}	35%
Pseudomonas aeruginosa	3.04×10^{-10}	30%
Streptococcus faecalis	5.47×10^{-10}	34%

* Based on Total Cell Count

ASSAYS ON URINE SPECIMENS

The amount of ATP per cell will affect the sensitivity of the assay as well as the accuracy for unknown samples. Environmental and metabolic factors are known to affect the ATP levels of bacteria. To evaluate this for urinary bacteria, the μg ATP per viable cell was measured as a function of growth phase in both trypticase soy broth and urine for *E. coli*.* When grown in urine, the ATP per cell was on the average two times higher than when grown in trypticase soy broth (figure 1), however, if glucose was added to urine or urea added to trypticase soy broth, the effect was reversed.

The centrifugation procedure was used to evaluate the correlation between colony counts and the ATP assay on infected urine specimens. On 183 urines routinely submitted to a bacteriology hospital laboratory, 0.1 ml of saline serial dilutions were plated on blood agar for colony counts. A scatter diagram of these points is given in figure 5, showing the femtograms of ATP per milliliter versus the colony forming units per milliliter.

The grid lines can be moved to indicate decision levels for calling a urine positive for infection or negative and reading its corresponding ATP level. The clustering at the extremes indicate either saturation or minimum sensitivity

---

* See Bush, V. N. et al., paper in this document.

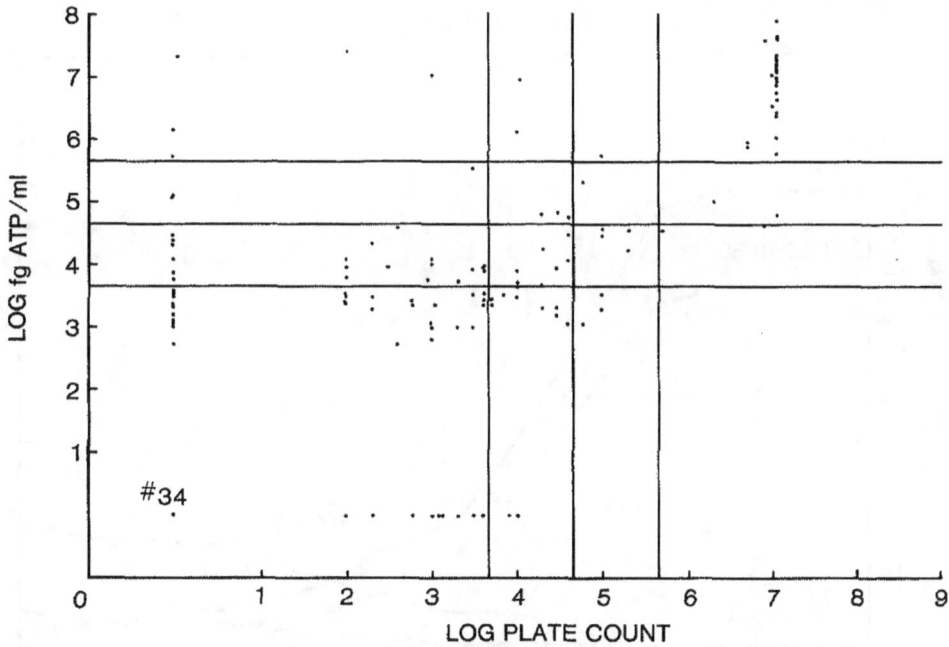

Figure 5. Correlation of ATP and colony counts in urine specimens.

levels for each of the measurement techniques. The #34 at a zero plate count and a zero ATP indicates that there were 34 urines that showed no colonies and blank level ATP measurement.

In order to determine whether a specimen is infected, a blank value is subtracted from each sample. A blank is obtained by using filtered, pooled urine and adding all the reagents used in the sample processing. We know, however, that the sample itself may affect the blank value and vary from sample to sample and therefore represents an uncertainty. In treating the data, therefore, subtract this blank value from each measurement, and then correlate these resulting values with bacteriologic counts. We can establish a number of light units which represents a lower limit around our blank value that gives the desired correlation, that is, choosing the trade-offs of false positives in deference to false negatives.

According to Kass (1957), 10^5 bacteria/ml indicates a urinary infection. When those specimens with a plate count below this level are considered to be negative, figure 6 shows the interpretation errors for a given ATP cutoff point when used to distinguish between ATP positive and ATP negative samples. The other two lines show this for a plate count infection level of 10^4/ml.

By grouping these data by categories representing a range of colony forming units per milliliter versus a range of femptograms of ATP per milliliter, table 7 shows the number of specimens that fell in each category and the percent of the total number of specimens.

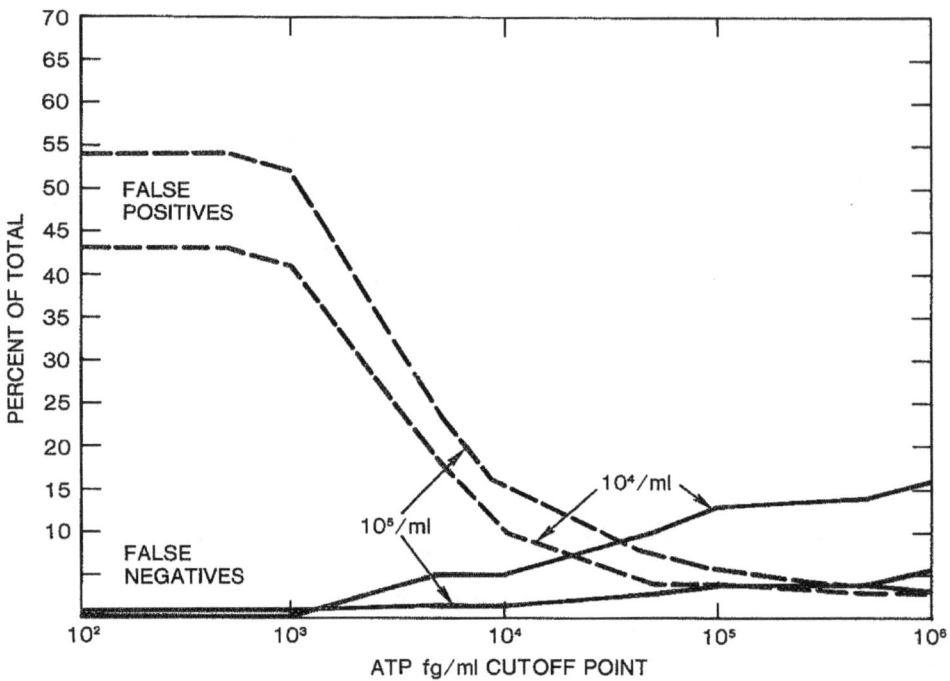

Figure 6. Malate buffer-nitric acid centrifugation procedure on urine specimens with a plate count greater than 10^4/ml and 10^5/ml.

Table 7

Number of Specimens and Percent of Total Number Tested from the Malate Buffer-nitric Acid Centrifugation Procedure on Urine Specimens

Plate Count	ATP (fg/ml)				
Colony Forming Units/ml	0 to 1.0×10^3	1.1×10^3 to 1.0×10^4	1.1×10^4 to 1.0×10^5	1.1×10^5 to 10^{10}	Total
0 – 1.0×10^3	(45)24.6%	(37)20.2%	(8)4.4%	(7) 3.8%	(97)53 %
1.1×10^3 – 1.0×10^4	(12) 6.6%	(19)10.4%	(0)0 %	(2) 1.1%	(33)18 %
1.1×10^4 – 1.0×10^5	(0) 0 %	(9) 4.9%	(9)4.9%	(3) 1.6%	(21)11.5%
1.1×10^5 – 10^{10}	(0) 0 %	(0) 0 %	(6)3.3%	(26)14.2%	(32)17.5%

ANAEROBIC COMPARISON

During the early trials of the ATP assay procedure it became apparent that roughly 10 percent of the assayed urines were negative by streak plate culture but positive by the ATP assay. However, in most of these cases microscopic counts showed better agreement with the ATP assay results. This can

22

be observed in the results shown in table 8. Thirty-three urine specimens were assayed for bacteria using the nonconcentrated ATP assay, 19 were strongly positive, and 14 were negative. The 19 positive specimens and 2 of the ATP negatives were cultured aerobically and anaerobically and counted microscopically. Five of the specimens were also tested for metabolic activity using BacTec (1972). The specimens were obtained directly from the patient clinics and anaerobic cultures immediately begun by Drs. Moore and Holdeman from the Virginia Polytechnical Institute's Anaerobe Laboratory. In many cases the anaerobic count was appreciably higher than the aerobic count and two cases which appeared negative via aerobic culture had at least 100,000 bacteria/ml when cultured anaerobically. Thus these results show that there are substantial numbers of organisms in urine that culture anaerobically and that in some cases these organisms do not culture aerobically, and the ATP assay does detect them.

Once it is realized that there can be substantial quantities of anaerobic bacteria in urine, the next problem is to determine the clinical significance of anaerobic bacteria in urine. Urine specimens were collected from 40 persons with no indication of urinary infection (midstream collection from 20 males and catheter collection from 20 females). Bacterial cell number estimates were made by aerobic and anaerobic cultures, microscopic counting, and an ATP assay. In every case but one, all the methods indicated that less than 1000 bacteria/ml were present. The one positive specimen was from a female and measured 1.8×10^8 cells/ml by anaerobic culture, 9.3×10^7 cells/ml by aerobic culture, and 4.8×10^8 cells/ml by ATP assay.

Thus, these two tests have indicated that there can be substantial amounts of anaerobic bacteria in urine with or without there being high numbers of aerobically culturable bacteria, and that anaerobic bacteria are not normally present in urine specimens from asymptomatic people. These are very preliminary results from a small number of samples and as such they are far from conclusive. However, the potential implication of the existence of fastidious anaerobic bacteria in some cases of urinary tract infection is substantial. It would be interesting to also determine anaerobic culture reports in a study such as that of Angell, Relman, and Robbins (1968).

This also indicates good agreement when used as a screening test on the general population, which is not expected to show many cases of positive urine culture.

SUMMARY

Efforts to develop a fast automatable system to detect the presence of bacteria in biological fluids, especially urine, have resulted in the optimization of procedures for use with different types of samples. These procedures have been validated by appropriate challenge systems in the laboratory and by the use of clinical specimens.

Table 8

Bacterial Cell Number Estimates by Various Methods

Number	ATP	Microscopic	Anaerobic	Aerobic	BacTec
1	2×10^6	2.4×10^4	2×10^4	1×10^5	ND*
2	4×10^5	1×10^5	4×10^3	2.3×10^5	ND*
3	1.7×10^6	1×10^5	1×10^2	1×10^2	Low+
4	5.7×10^8	2×10^7	6×10^7	Neg	Neg
5	5.6×10^5	2×10^4	1×10^2	Neg	ND*
6	5.7×10^7	6×10^7	1×10^8	4.4×10^7	High+
7	Neg	4×10^4	6×10^2	Neg	ND*
8	1.6×10^5	2×10^4	1×10^2	1×10^2	ND*
9	2.6×10^6	2×10^4	1×10^2	1×10^3	Neg
10	6×10^5	8×10^5	2×10^6	7×10^3	ND*
11	2×10^5	4×10^4	1×10^2	Neg	ND*
12	Neg	4×10^4	2×10^5	3×10^3	ND*
13	3.4×10^6	1×10^7	2×10^6	6.5×10^3	V.Low+
14	1.3×10^5	1×10^5	1×10^4	5×10^3	ND*
15	1×10^5	4×10^3	1×10^5	Neg	ND*
16	4×10^7	2×10^8	4×10^8	10^8	ND*
17	6×10^6	3×10^6	4×10^5	1.3×10^4	ND*
18	3×10^7	2×10^7	5×10^6	3.4×10^6	ND*
19	1.8×10^6	2×10^4	1×10^2	Neg	ND*
20	1.3×10^6	2×10^6	1×10^6	8×10^6	ND*
21	9×10^5	4×10^4	4×10^2	Neg	ND*
22†					

* ND = Not Done

† No. 22 through No. 33 were negative by all methods.

Improvements in reproducibility and accuracy by reducing interfering substances have resulted, as well as increase in sensitivity by decreasing the background and concentrating the sample.

The procedure for removing up to 10^6 leucocytes/ml of urine and measuring the bacterial ATP from as few as 1000 urinary pathogens/ml of urine when starting with a 10-ml sample and concentrating by centrifugation has been developed and used on a small number of clinical specimens. Correlation with culture procedures shows that the luciferase assay can be used to quantitate bacteria from urine cultures with low percentages of false positives and false negatives.

Further efforts will include evaluation with several thousand clinical specimens and correlation with additional types of measurements, such as mini-culture methods, metabolic measurements, and anaerobic culture.

ACKNOWLEDGMENT

This work was supported in part by Technology Applications and Technology Utilization of NASA and by the Regional Medical Program Service, DHEW.

REFERENCES

Angell, M. E., A. S. Relman, and S. L. Robbins, "Active Chronic Pyelonephritis Without Evidence of Bacterial Infection," *NEJM,* **278**, 1968, p. 1303.

BacTec, Johnston Laboratories, Inc., Cockeysville, Md. 21030, USA Patent No. 3,676,679, 1972.

Chappelle, E. W., Method for the Detection of Viruses. U.S. Patent No. 3,575,812, 1971.

Chappelle, E. W. and G. V. Levin, "The Design and Fabrication of an Instrument for the Detection of Adenosine Triphosphate (ATP)," NASA CR-411, 1966.

Chappelle, E. W. and G. V. Levin, "The Use of the Firefly Bioluminescent Assay for the Rapid Detection and Counting of Bacteria," *Biochem. Med.,* 2, 1968, p. 49.

D'Eustachio, A. J. and G. V. Levin, Abst. *Proc. Am. Soc. of Microb.,* **144**, 1967.

Harvey, E. N., *Living Light,* Hafner Publishing Co., Inc., 1965.

Kass, E. H., "Bacteriuria and the Diagnosis of Infections of the Urinary Tract," *Arch. Int. Med.,* **100**, 1957, p. 709.

Klofat, W., G. L. Picciolo, E. W. Chappelle, and E. Freese, "Production of Adenosine Triphosphate in Normal Cells and Sporulation Mutants of *Bacillus subitilis*," *J. Bio. Chem.,* **244**, 1969, p. 3270.

Levin, G. V., J. R. Clendenning, E. W. Chappelle, and E. Rocek, "Rapid Method for Detection of Microorganisms by ATP Assay; Its Possible Application in Virus and Cancer Studies," *BioScience,* **14**, 1964, p. 37.

McElroy, W. D., and B. Glass, eds., "A Symposium on Light and Life," sponsored by Johns Hopkins University, 1960, Baltimore, The J. H. Press, 1961.

McElroy, W. D., H. H. Selinger, and E. H. White, "Mechanism of Bioluminescence, Chemiluminescence, and Enzyme Function in the Oxidation of Firefly Luciferin," *Photochem. Photobiol.,* **10**, 1969, p. 153-170.

Picciolo, G. L., B. N. Kelbaugh, E. W. Chappelle, and A. J. Fleig, "An Automated Luciferase Assay of Bacteria in Urine," NASA TM X-65521, 1971.

Plant, P. J., E. H. White, and W. D. McElroy, "The Decarboxylation of Luciferin in Firefly Bioluminescence," *Biochem. Biophy. Res. Comm.,* **31**, 1968, p. 98-103.

St. John, J. B., "Determination of ATP in *Chlorella* with the Luciferin-Luciferase Enzyme System," *Analytical Biochem.,* **37**, 1970, p. 409.

Strehler, B. L., "Methods of Enzymatic Analysis," H. -U. Bergmeyer, ed., 2nd ed., Academic Press, 1965, p. 559.

Vlodavsky, I., M. Inbar, and L. Sachs, "Membrane Changes and Adenosine Triphosphate Content in Normal and Malignant Transformed Cells," *Proc. Nat. Acad. Sci. USA,* **70**, (6), June 1973, p. 1780-1784.

A COMPARISON OF CERTAIN EXTRACTING AGENTS FOR EXTRACTION OF ADENOSINE TRIPHOSPHATE (ATP) FROM MICROORGANISMS FOR USE IN THE FIREFLY LUCIFERASE ATP ASSAY

Elizabeth A. Knust
New England Medical Center
Boston, Massachusetts

Emmett W. Chappelle and Grace Lee Picciolo
Goddard Space Flight Center
Greenbelt, Maryland

The importance of complete extraction of adenosine triphosphate (ATP) and the ability to assay with minimal inhibition led to a comparison of different ATP extracting agents for use in the firefly luciferase ATP assay. This assay can be used in clinical and industrial applications, such as determination of urinary infection levels, microbial susceptibility testing, and monitoring of yeast levels in beverages.

The optimal extracting agent is one which provides maximal extraction of ATP and minimal inhibition of the luciferase enzyme. Three categories of extractants were investigated for their extracting efficiency. They were ionizing organic solvents, nonionizing organic solvents, and inorganic acids. To represent the ionizing organic solvents, dimethylsulfoxide (DMSO) (Chappelle and Levin, 1964) and formamide were used. For the nonionizing organic solvents n-butanol (Chappelle and Levin, 1968), chloroform,* ethanol (St. John, 1970), methanol (St. John, 1970), acetone (Chappelle and Levin), and methylene chloride (dichloro-methane) were used. And finally, for the inorganic acids category, nitric acid† and perchloric acid (Picciolo et al., 1971) were chosen. Concentrations used are given in the list below. The references cited for the above agents are for the agent as an ATP extractant, and the procedure used in the reference is not necessarily the one used in this study. These extracting agents were used on certain urinary tract pathogenic bacteria and yeast. They were also used on *Saccharomyces carlsbergensis* (Brewer's yeast).

---

*Dhople, A.M., J.H. Hanks, and E.W. Chappelle, "Ultrasensitive Method for Detection of Microbial and Mycobacter ATP," *Proc. Amer. Soc. Microbiol. Ann. Meeting*, 1971.
†See Picciolo et al., paper in this document.

Concentration of Extractants
Used in Comparative Study

Chloroform 100%
Methanol 100%
Ethanol 100%
DMSO 30%
Formamide 10%
N-Butanol 6%
Methylene Chloride 90%
Acetone 90%
PCA 0.1 N and 1.0 N
HNO_3 0.1 N and 1.0 N

The urinary tract pathogens used in the study were obtained from a clinical laboratory, and the *Saccharomyces carlsbergensis* used was obtained from a brewery. The urinary tract pathogens consisted of *Escherichia coli, Staphylococcus aureus, Klebsiella pneumoniae, Enterobacter species, Proteus mirabilis, Proteus vulgaris, Staphylococcus epidermidis, Streptococcus faecalis, Pseudomonas aeruginosa,* and *Candida albicans.* These were grown at 310 K (37°C) for 16 to 18 hours in trypticase soy broth. The *Saccharomyces* was grown in wort broth for 40 to 42 hours at 310 K (37°C). The bacteria were grown while shaking; the yeast were not. The organisms were centrifuged at 10,500 RCF $\times$ G for 5 minutes and the supernatant decanted. The organisms were then treated with the respective extracting reagent in its optimal extracting condition, with the optimal condition having been determined prior to this.

The procedure for each extracting agent was then followed, and the final diluent water was added. The sample was then assayed on the DuPont Biometer using DuPont firefly luciferase-luciferin, reconstituted in TRIS buffer at an optimal concentration and pH complimentary to the respective extracting agents.

The light units of the sample and the light units of the ATP standard that was run with each extracting agent were used to calculate micrograms per milliliter of ATP extracted. These results were then compared for extraction efficiency.

In comparing the organic extractants, the acetone extracted more ATP per milliliter of bacteria than the other organic extractants. It was noted that among the extractants, inhibition of the luciferase was not present with acetone or methylene chloride because these volatile solvents were boiled off.

Figure 1 shows the relative extraction efficiency of nitric acid (HNO_3), acetone, dichloromethane, n-butanol, formamide, and DMSO on *Pseudomonas aeruginosa,* a gram-negative organism. Figure 2 shows the same extractants on *Staphylococcus aureus,* a gram-positive organism.

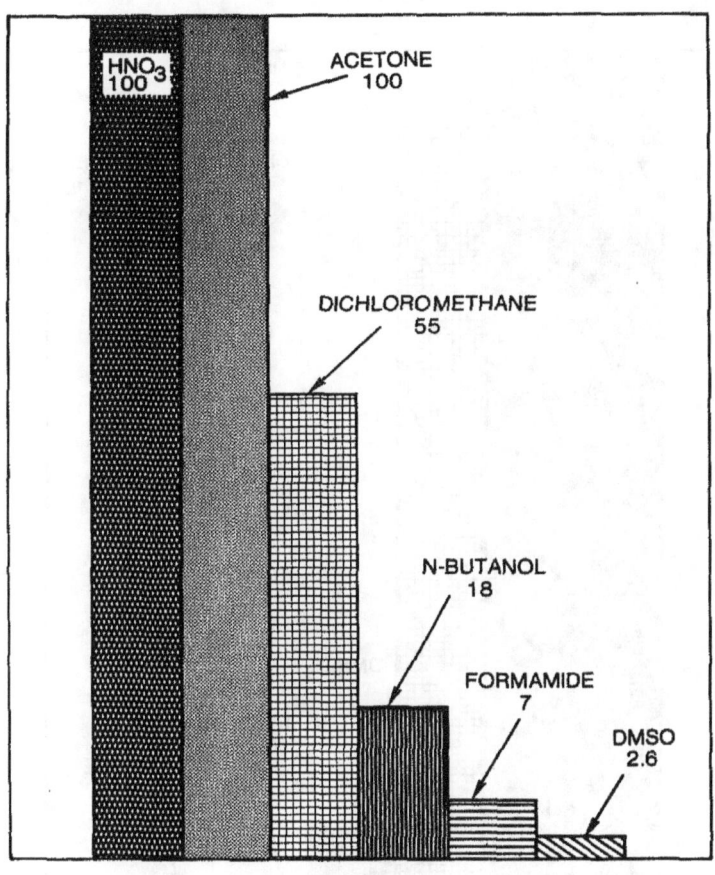

Figure 1. Relative extraction efficiency of various agents on *Pseudomonas aeruginosa.*

In figure 3 is the relative extraction efficiency of nitric acid, acetone, methanol, ethanol, and chloroform on *Klebsiella pneumoniae,* a gram-negative organism.

Figure 4 shows a comparison of acetone and chloroform on *Escherichia coli, Klebsiella pneumoniae, Proteus mirabilis,* and *Pseudomonas aeruginosa,* all of which are gram-negative, and on *Streptococcus faecalis,* a gram-positive organism.

Nitric and perchloric acid were found to be comparable in extraction efficiency. The acetone was then compared to the inorganic acid extractants after each procedure had been optimized for both bacterial and yeast ATP extraction.

The results obtained with inorganic acids and the acetone were comparable in extraction efficiency. It was noted that in the two procedures there was a 2 to 5 percent variation in injection values. This would result in the inorganic acid ATP extraction value and acetone ATP extraction value varying 2 to 5 percent.

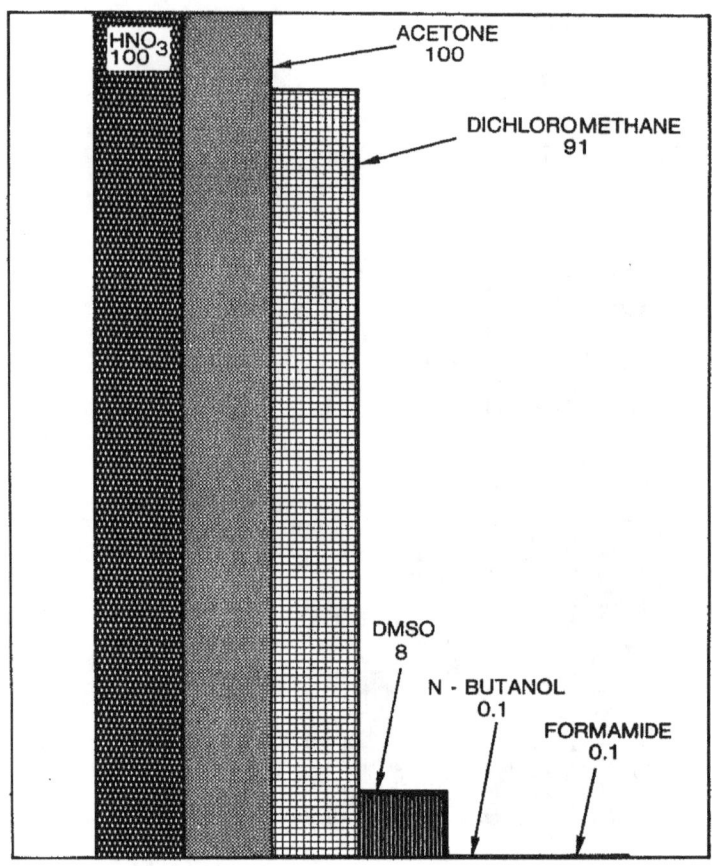

Figure 2. Relative extraction efficiency of various agents on *Staphylococcus aureus*.

In procedure 1, the acetone procedure for both bacterial and yeast ATP extraction is listed. Ten milliliters of sample are centrifuged and the supernatant decanted. The acetone is then added and the sample heated to permit the volatile solvent to boil off. The 0.5 milliliter nonvolatilized is then assayed with DuPont luciferase-luciferin on the DuPont Biometer.

Procedure 2 shows the nitric acid procedure for bacterial ATP extraction. Ten milliliters of sample are centrifuged and the supernatant decanted. The HNO_3 acid is added and, after 5 minutes, the diluent is added. The 0.4 milliliter is then assayed with DuPont luciferase-luciferin on the DuPont Biometer.

It was also noted that the blank value obtained with acetone was much lower than the blank value obtained with nitric acid and perchloric acid.

CONCLUSIONS

With results obtained in the study, it was evident that the acetone can also be used as a bacterial and yeast ATP extractant comparable to the inorganic acids.

30

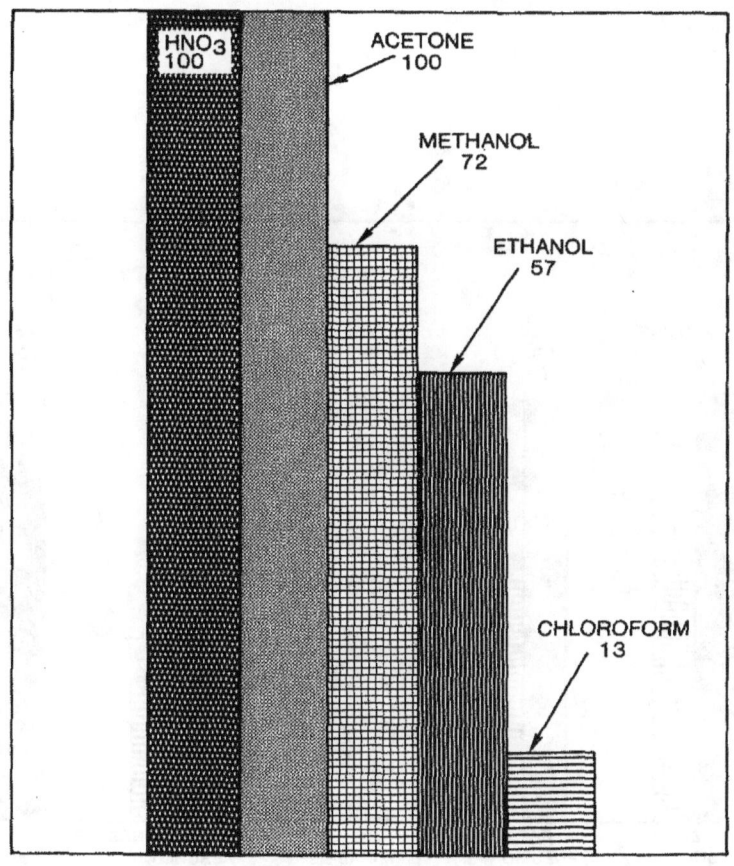

Figure 3. Relative extraction efficiency of various agents
on *Klebsiella pneumoniae.*

There were several advantages to the acetone procedure. One of the most important advantages became evident with the yeast. In using nitric acid or perchloric acid it would require an initial acid concentration of 1.0 N to completely extract the ATP from the yeast. The result of using this concentration of acid required a dilution factor of 10 to allow for assay with the luciferase enzyme uninhibited. This dilution factor resulted in a loss of sensitivity. The acetone procedure was adequate for extracting bacteria or yeast with no adjustment needed.

Another advantage in the acetone procedure was the blank value obtained. The lower value obtained with the acetone procedure would allow for a wider range of ATP to be measured, thus lending to the ability to detect fewer microorganisms.

It was also shown that the advantage of having the acetone boiled off left no reagent to inhibit the luciferase and allowed for a lower molarity of TRIS buffer to be used which would increase luciferase activity.

31

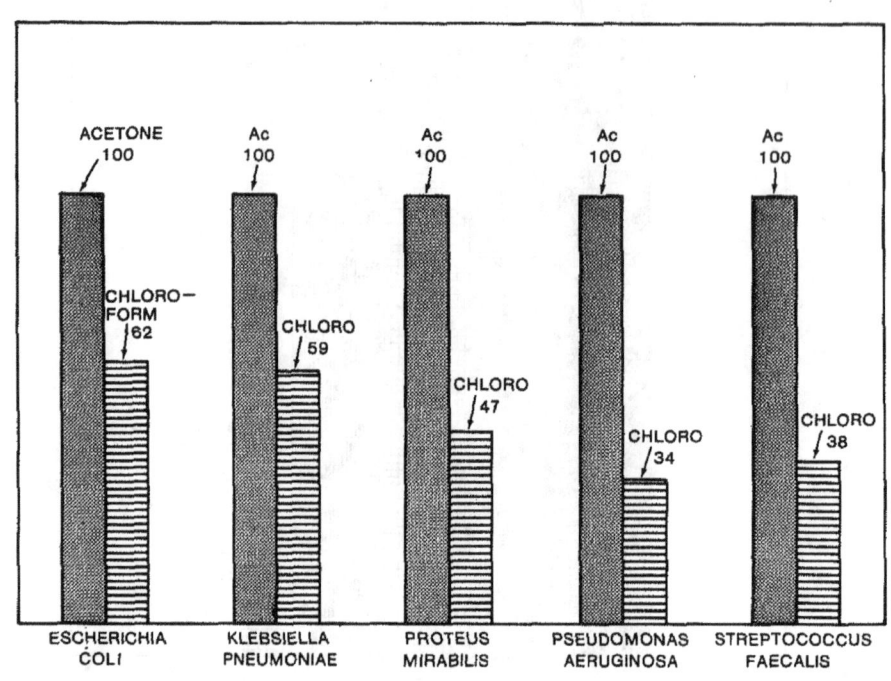

Figure 4. Relative extraction efficiency of acetone and chloroform on various bacteria.

Procedure 1
Acetone Procedure for Extraction of Yeast and Bacteria

1. 10 ml sample.

2. Centrifuge 10,500 RCF X G 5 minutes.

3. Decant supernatant 5 minutes.

4. Add 5.0 ml, 90 percent acetone (diluted with H_2O)—vortex.

5. Heat 40 minutes at 363 K (90°C) (until odor of acetone not present).

6. Let cool—vortex (0.5 ml should be remaining in tube after acetone has boiled off).

7. Assay: Inject 0.1 ml of above into 0.1 ml luciferase reconstituted with 1.5 ml of 0.05 M TRIS with 0.01 M $MgSO_4$ pH 7.75 per vial DuPont luciferase.

< not applicable>

Procedure 1 (Continued)

8. An ATP standard of 1.0 μg/ml or 10^{-1} μg/ml should be used. Use 0.05 ml of ATP standard plus 5.0 ml of acetone and heat. Then assay.

9. Blanks should be run on the media in which the bacteria are suspended.

10. Calculation of micrograms ATP/ml should be done with consideration for the final volume of sample (0.50 ml) and for the ATP standard (0.55 ml).

Procedure 2
HNO$_3$ Procedure for Extraction of Bacteria

1. 10 ml sample.

2. Centrifuge 10,500 RCF $\times$ G 5 minutes.

3. Decant supernatant 5 minutes.

4. Add 0.2 ml, 0.1 N HNO$_3$.

5. Wait 5 minutes.

6. Add 0.2 ml of sterile, distilled, deionized water—vortex well.

7. Assay: Inject 0.1 ml of above into 0.1 ml luciferase reconstituted with 1.5 ml of 0.20 M TRIS with 0.01 M MgSO$_4$ pH 8.3 per vial DuPont luciferase.

8. An ATP standard of 1.0 μg/ml or 10^{-1} μg/ml should be used. 0.05 ml of ATP standard plus 0.2 ml 0.1 N HNO$_3$, then add 0.2 ml of HOH and assay.

9. Blanks should be run on the water used and the media the bacteria are suspended in.

10. Calculation of micrograms ATP/ml should be done with consideration for the final volume of sample (0.40 ml) and for the ATP standard (0.45 ml).

There are also disadvantages to the acetone procedure. The most evident one is the 40-minute heating time as compared to the nitric acid and perchloric acid extraction time of 5 minutes. It was felt that the flammability and requirement of a hood for safe ventilation of the fumes were also disadvantages.

Over all, the acetone procedure is equivalent to the inorganic acid procedure in ATP extraction efficiency and can be used in applicable situations to an advantage.

REFERENCES

Chappelle, E. W. and G. V. Levin, "The Use of the Firefly Bioluminescent Assay for the Rapid Detection and Counting of Bacteria," *Biochem. Med.*, 2, 1968, p. 49.

Chappelle, E. W. and G. V. Levin, "Rapid Microbiological Detection," Navy Contractor Report 178-8097, 1964.

Picciolo, G. L., B. N. Kelbaugh, E. W. Chappelle, and A. J. Fleig, "An Automated Luciferase Assay of Bacteria in Urine," NASA TM X-65521, 1971. (Also in *Proc. Chemical Abs.*)

St. John, J. B., "Determination of ATP in Chlorella with the Luciferin Luciferase Enzyme System," *Analytical Biochem.*, 37, 1970, p. 409.

THE EFFECT OF GROWTH PHASE AND MEDIUM ON THE USE OF THE FIREFLY ADENOSINE TRIPHOSPHATE (ATP) ASSAY FOR THE QUANTITATION OF BACTERIA

V. N. Bush
Department of Biology
Delaware State College
Dover, Delaware

Grace Lee Picciolo and Emmett W. Chappelle
Goddard Space Flight Center
Greenbelt, Maryland

ABSTRACT

The luciferase assay for adenosine triphosphate (ATP) has been suggested for use as a rapid method to determine the number of bacteria in a urine sample after nonbacterial components of the urine are removed. Accurate cellular ATP values, determined when bacteria are grown in an environment similar to that in which they are found in urine, are necessary for the calculation of bacterial titer in urine. Cellular ATP values vary depending on the method of extraction, the growth phase of the cells, and the growth conditions of the cells. ATP per cell values of stationary phase *E. coli* grown in urine were two times greater than ATP per cell values of cells grown in trypticase soy broth. Glucose and urea were examined as possible components responsible for the cellular ATP variation. Glucose was added to sterile glucose-free pooled urine, and urea was added to trypticase soy broth. The cells in which urea was added to trypticase soy broth had ATP per cell values similar to those found when cells were grown in urine.

The luciferase assay for adenosine triphosphate (ATP) has been extensively applied to areas of bacterial detection. In this work, an *E. coli* population in different physiological states and grown in different media was examined to determine variations in ATP per viable cell values. This work was done to determine ATP per cell values of bacterial cells obtained from clinical urine samples which would reflect the physiological state of the cells at the time when the luciferase ATP assay would be done.

All work was done using a clinical urinary tract isolate of *E. coli*. *E. coli* was selected for use because it is the cause of the majority of urinary tract infections.

A logarithmically growing culture of *E. coli* in trypticase soy broth was added to two growth media: (1) trypticase soy broth, a general purpose medium, and (2) urine. The urine was collected at random from healthy male and female adults. All samples were initially tested with Uristix and were rejected if glucose or protein was detected using this measurement. These samples were usually the second-morning specimens. The urine was filtered through a 0.22-micrometer Millipore filter after prior filtration which removed larger particulate material in the urine. *E. coli* was grown in the two media at 310 K (37°C) without shaking. At periodic intervals, the population was analyzed for intracellular ATP content and the number of viable *E. coli* present.

Procedure 1 shows the procedure used to extract ATP from *E. coli*. The ATP from the bacterial cells was extracted using nitric acid, after a prior treatment with apyrase, an ATPase, to remove any nonbacterial ATP present.

Procedure 1
Procedure to Extract ATP from Bacteria

1. 0.5 ml bacterial cells with their growth medium +0.1 ml 40 mg apyrase/ml 0.03 M $CaCl_2$ are mixed and allowed to sit for 15 minutes.

2. Add 0.1 ml 1.5 N HNO_3.
   Allow to sit 5 minutes.

3. Volume brought to 5 ml with deionized H_2O.

Procedure 2 shows the procedure used to treat standard ATP. Instead of using the bacterial cells and their growth medium as the starting sample, the bacteria were filtered out of the growth medium using a 0.22-micrometer Millipore filter. The filtrate was treated with apyrase and nitric acid in the same manner as the bacterial sample. A known amount of ATP was then added to the reaction mixture.

Procedure 2
Procedure to Treat Standard ATP

1. 0.5 ml growth medium (bacterial cells removed) +0.1 ml 40 mg apyrase/ml 0.03 M $CaCl_2$. Mix. Wait 15 minutes.

2. Add 0.1 ml 1.5 N HNO_3. Mix. Wait 5 minutes.

3. Add 0.1 ml ATP (1 μg/ml).

4. Volume brought to 5 ml with deionized H_2O.

The procedure to treat the blank to extract any ATP is shown in procedure 3. The growth medium from which bacterial cells were filtered was treated with apyrase and nitric acid.

Procedure 3
Procedure to Treat Blank to Extract ATP

1. 0.5 ml growth medium (bacterial cells removed) +0.1 ml 40 mg apyrase/ml 0.03 M $CaCl_2$. Mix. Wait 15 minutes.

2. Add 0.1 ml 1.5 N HNO_3. Mix. Wait 5 minutes.

3. Volume brought to 5 ml with deionized H_2O.

After the ATP was extracted from the bacterial sample, standard ATP, and blank, the assay to measure the amount of ATP was performed as follows. One-tenth milliliter of the treated sample was injected by needle and syringe into 0.1 ml of rehydrated DuPont luciferase (3 ml 0.2 M TRIS, 0.01 M $MgSO_4$, pH 8.25 per vial). The luciferase was located in a cuvette in the light-tight chamber of the Chem-Glow (Aminco) instrument, which was attached to an X-Y recorder (Hewlett-Packard) to obtain a permanent record of the amount of ATP present in each sample. Using the blank, standard, and sample, the amount of ATP per milliliter in the bacterial sample was determined.

The number of viable cells was determined by pour plating in which serial dilutions of the population were made and then duplicate 0.1-ml aliquots of each dilution were added to 18 ml of melted trypticase soy (TS) agar. The plates were incubated at 310 K (37°C) for 24 hours, and the total number of colonies was counted. The number of viable cells per milliliter was then calculated.

Figure 1 shows the ATP per viable cell values when *E. coli* was grown in TS and urine for various time lengths. The growth curves of *E. coli* in TS and urine during the logarithmic phase are very similar. The number of cells that each medium will support during the stationary phase is about 1.2×10^9 cells per ml in TS broth and 4×10^8 cells per ml in urine. This pooled urine is a good growth medium for *E. coli*; the factors limiting the growth of *E. coli* in urine as compared to the growth which can be obtained in TS broth were not investigated.

Examination of the ATP per viable cell value of *E. coli* grown in both urine and TS broth shows a reduction in value as the cells grow logarithmically. The lowest ATP per cell values of *E. coli* grown in both urine and TS broth occur during the stationary phases of the populations.

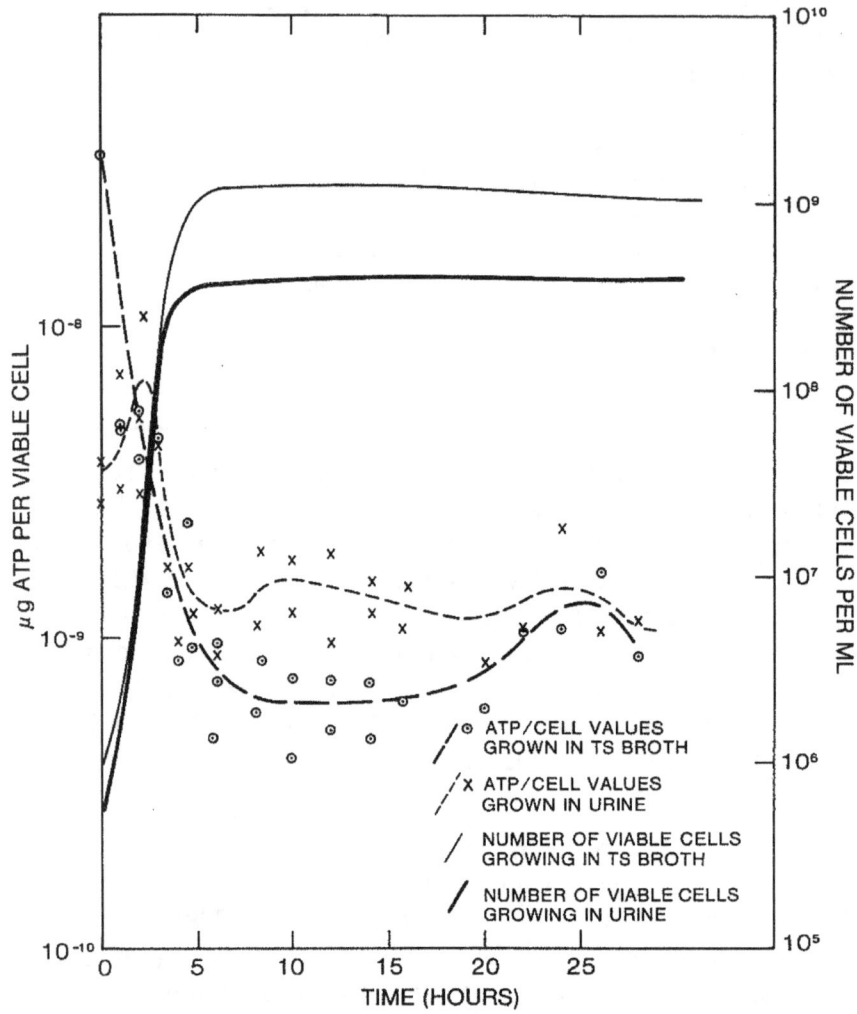

Figure 1. Effect of growth medium and the length of time grown in this medium on ATP per viable *E. coli* values.

In comparing the ATP per cell values of stationary phase populations of *E. coli* in TS and urine, a two-fold difference is found between the cells grown in these two media. The average amount of ATP per viable cell of cells grown in urine is approximately 1.2×10^{-9} µg ATP per viable cell and 8×10^{-10} µg ATP per viable cell when grown in TS broth. The mean difference between TS and urine ATP per viable cell values when the two populations are in stationary phase is 4.88×10^{-10} µg ATP per cell.

Our next investigation was to determine the cause of the large difference in ATP per cell values found in the stationary phase between urine and TS grown *E. coli* populations. Possible explanations for this phenomenon were (1) a growth factor contained in urine or TS that is unique to that medium or (2) a metabolite produced as a consequence of growth that is affecting the ATP per viable cell value. We decided to examine two factors,

38

glucose and urea, to see their effects on the ATP per viable cell values of stationary phase cultures of *E. coli.* The urine we used had no detectable glucose in it as measured by the Uristix method which can detect 0.1 percent glucose levels. TS broth had a glucose content of 2.5 g per liter. We added various concentrations of glucose to the urine prior to the addition of *E. coli.* The cells were then grown in the same manner as previously described and the same measurements made on stationary phase cells. The effect of glucose addition to urine on ATP per viable cell values of stationary *E. coli* is shown in figure 2. As more glucose was added to urine, there was a decrease in ATP per viable cell values. When the glucose concentration in urine was similar to the glucose concentration in TS broth, there was very little difference in ATP per cell values when cells were grown under these two conditions.

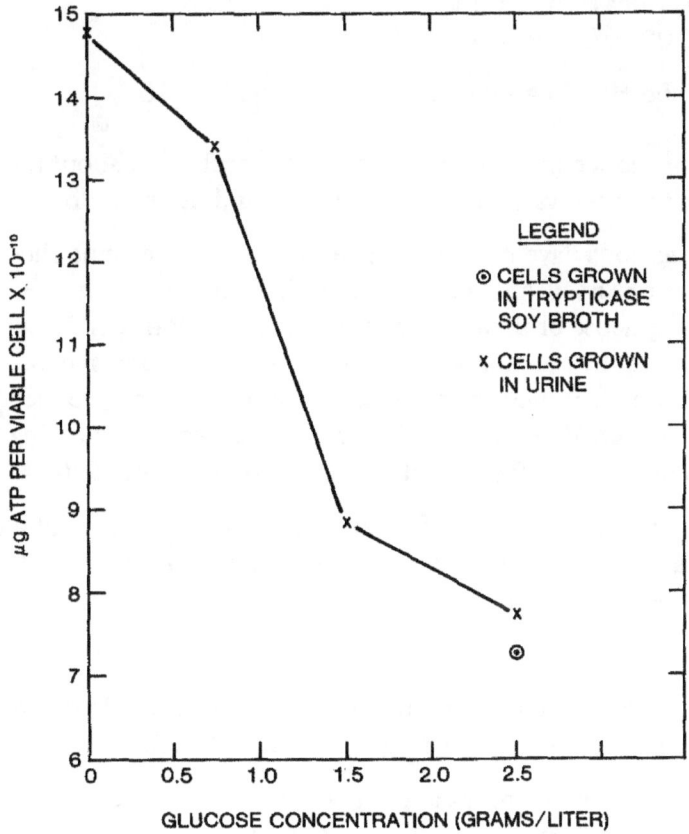

Figure 2. The effect of glucose addition to urine on ATP per viable cell values of stationary *E. coli.*

In investigating the effect of urea on ATP per viable cell values, we added urea to trypticase soy broth so that its concentration would be comparable to that found in urine. *E. coli* was inoculated into these growth media and grown for 16 to 18 hours at 310 K (37°C). The addition of urea to TS broth caused an increase in ATP per cell value to a level higher than that found in cells grown in urine as shown below.

Effect of Urea on ATP per Viable Cell Values of a Stationary Phase Population of *E. Coli*

Growth Medium	μg ATP per Viable Cell
Trypticase soy broth	5.29×10^{-10}
Urine	9.14×10^{-10}
Trypticase soy broth +0.025 g/ml urea	1.22×10^{-9}
Urine +0.025 g/ml urea	1.76×10^{-9}

Adding the same concentration of urea to urine resulted in about a two-fold increase in ATP per cell value above the value found in urine alone.

Further investigations have not been made to date to determine the explanation for these observed ATP per cell value differences. It may be that the presence of glucose or urea may be the factor regulating ATP per cell values or there may be a metabolite produced when glucose is metabolized by *E. coli* or a chemical such as ammonia or carbon dioxide produced during the spontaneous breakdown of urea in the medium that is regulating intracellular ATP values (Hempfling, 1970; Pastan and Perlman, 1970).

In summary, if use is to be made of a determined ATP per cell value in calculating the number of bacteria present in a sample such as a clinical urine sample, this determined ATP per cell value must take into account the following factors:

- The bacteria found as urinary tract pathogens should be grown under various conditions to get a range in ATP per cell values.

- Temperature, pH, nutrients in the medium, and oxygen tension are a few factors which must be considered (Forrest, 1965; Cole et al., 1967; Strange et al., 1963; Bailey and Parks, 1972).

As shown in this work and by previous investigators (Hamilton and Holm-Hansen, 1967), the age of the population makes a difference in the amount of ATP found in each cell. It is also important that the same procedure that will be used to treat the sample (such as urine) is followed to arrive at ATP per cell values.

REFERENCES

Bailey, R. B. and L. W. Parks, "Response of the Intracellular ATP Pool of *Saccharomyces cerevisiae* to Growth Inhibition Induced by Excess L-Methionine," *J. Bact.*, **111**, 1972, p. 542-546.

Cole, H.A., J.W.T. Wimpenny, and D.E. Hughes, "The ATP Pool in *E. coli*. I. Measurement of the Pool Using a Modified Luciferase Assay," *Biochim. Biophys. Acta*, **143**, 1967, p. 445-453.

Forrest, W.W., "Adenosine Triphosphate Pool During the Growth Cycle in *Streptococcus faecalis*," *J. Bact.*, **90**, 1965, p. 1013-1016.

Hamilton, R.D. and O. Holm-Hansen, "Adenosine Triphosphate Content of Marine Bacteria," *Limnology and Oceanography*, **12**, 1967, p. 319-324.

Hempfling, W.P., "Repression of Oxidative Phosphorylation in *E. coli* by Growth in Glucose and Other Carbohydrates," *Biochem. Biophys. Res. Comm.*, **41**, 1970, p. 9-15.

Pastan, I. and R. Perlman, "Cyclic Adenosine Monophosphate in Bacteria," *Science*, **169**, 1970, p. 339-344.

Strange, R.E., H.E. Wade, and F.A. Dark, "Effect of Starvation on Adenosine Triphosphate Concentration in *Aerobacter aerogenes*," *Nature*, **199**, 1963, p. 55-57.

A RAPID METHOD FOR THE DETERMINATION OF MICROBIAL SUSCEPTIBILITY USING THE FIREFLY LUCIFERASE ASSAY FOR ADENOSINE TRIPHOSPHATE (ATP)

H. Vellend, S. A. Tuttle, M. Barza, and L. Weinstein
New England Medical Center Hospital
Boston, Massachusetts

Grace Lee Picciolo and Emmett W. Chappelle
Goddard Space Flight Center
Greenbelt, Maryland

ABSTRACT

The luciferase assay for adenosine triphosphate (ATP) was optimized for pure bacteria in broth in order to evaluate if changes in bacterial ATP content could be used as a rapid measure of antibiotic effect on microorganisms.

Broth cultures of log phase bacteria (10^6 colony-forming units/ml) were incubated at 310 K (37°C) for 2.5 hours at antimicrobial concentrations which resulted in the best discrimination between "sensitive" and "resistant" strains. ATP assays were performed on the control broth culture at the onset of incubation (A_0) and again after 2.5 hours on both the antibiotic-containing broth culture (B_t) and the antibiotic free growth control (A_t). The drug effect on bacterial ATP content was quantitated using the following formula:

$$\text{ATP Index} = \frac{\log B_t - \log A_0}{\log A_t - \log A_0}$$

Empirical observations from a large number of microbial susceptibility tests performed by this method suggested that an index of $> +0.25$ implied resistance and $\leq +0.25$, sensitivity.

Eighty-seven strains of 11 bacterial species were studied for their susceptibility to 12 commonly used antimicrobial agents: ampicillin, Penicillin G, nafcillin, carbenicillin, cephalothin, tetracycline, erythromycin, clindamycin, gentamicin, nitrofurantoin, colistin, and chloramphenicol. An overall comparison of the results obtained by the ATP index and agar diffusion sensitivity testing demonstrated a 90 percent agreement. Of the 10 percent instances of disagreement, most (three-fourths) were major, i.e., false-resistance or false-sensitivity. One-quarter of the disagreements were minor, i.e., intermediate by agar diffusion and either sensitive or resistant by the ATP index. The principal cause for major disagreement between the ATP index and agar diffusion appeared to be related to the mode of action of the antimicrobial agent. The reproducibility of the method was entirely satisfactory (94 percent).

The major advantage of the ATP system over existing methods of rapid microbial susceptibility testing is that the assay can be made specific for bacterial ATP. This unique feature may allow this technique to be applied directly to organisms in urine or other biological fluids without prior bacterial isolation or subculture. Studies to this effect are in progress.

APPLICATION OF THE LUCIFERIN-LUCIFERASE ENZYME SYSTEM FOR DETERMINATION OF ADENOSINE TRIPHOSPHATE (ATP) TO STUDIES ON THE MECHANISMS OF HERBICIDE ACTION

J. B. St. John
Agricultural Research Service
U.S. Department of Agriculture
Beltsville, Maryland

Adenosine triphosphate (ATP) pivotally supplies energy for biosynthetic reactions. Photosynthetic and oxidative phosphorylation are the two major processes through which a chlorophyllous organism produces ATP. Despite many reports of herbicidal effects on these processes in vitro, much less is known about herbicidal effects on phosphorylation processes in vivo. We have used the firefly luciferin-luciferase assay, which measures ATP directly, to study herbicide effects on phosphorylation processes in vivo. A detailed study of our method, as well as the use of the Aminco Chem-Glow Photometer,* has been published elsewhere (St. John, 1970). Briefly, our system is as follows:

1. ATP is extracted by boiling for 1 minute in a suitable solvent. We have found absolute ethanol or water or a combination of the two to be most suitable for algae and other plant materials.

2. The extracts are blown dry with nitrogen and reconstituted with water.

3. A 1-ml fraction is diluted to 2 ml and made 0.025 M in HEPES-Mg^{++} buffer, pH 7.5.

4. A second 1-ml fraction is treated similarly, but a known amount of ATP is added.

5. By use of an Aminco Chem-Glow Photometer attached to an integrator-timer, light emission is integrated for 30 seconds after the mixing of a 0.45-ml sample with 0.1-ml reconstituted, commercially available, firefly lantern extract.

---

* Mention of a trade name or proprietary product does not constitute a guarantee or warranty of the product by the U.S. Department of Agriculture and does not imply its approval to the exclusion of other products that may also be suitable.

6.   ATP concentration is determined by solution of this equation:

$$C\mu = \frac{I\mu\, C_s\, D_f}{I\mu + C - I\mu} \qquad (1)$$

Here $C\mu$ is the ATP concentration in moles per liter, $I\mu$ is the relative integrated intensity produced in response to the unknown concentration of ATP, C_s is the known concentration of ATP in moles per liter, D_f is a dilution factor, and $I\mu + C$ is the relative integrated intensity produced in response to the unknown plus the known concentration of ATP.

This method has several advantages. The method of constant addition for quantitation of ATP successfully overcomes interference caused by substances normally found in biological extracts, the added herbicides, the solvent used for extraction, and such added things as ATP, AMP, GTP, and PP. Thirty-second integration of the light emitted after the mixing of the sample and enzyme, which includes the peak intensity of the initial flash and a portion of the decaying light,

- Minimizes the variables of injection and sample mixing;

- Is highly reproducible, with less than a 2 percent relative standard deviation;

- Affords a level of sensitivity equal to that obtained with purified luciferin-luciferase preparations—namely, 10^{-9} M ATP solutions or picomole quantities of ATP with linearity extending over at least a thousand-fold range; and

- Costs about 3 cents a test as compared with 30 cents a test when purified luciferin-luciferase preparations are required.

To evaluate herbicide effects on phosphorylation processes in vivo, we selected *Chlorella* as the test organism and diuron [3 – (3,4-dichlorophenyl) –1, 1-dimethylurea] and chlorpropham (isopropyl *m*-chlorocarbanilate) as test chemicals (St. John, 1971). When *Chlorella* cells were grown in the light on a totally inorganic medium, conditions that favor photosynthetic phosphorylation as the major system for ATP production, both diuron and chlorpropham reduced ATP levels and growth. These findings indicated that both diuron and chlorpropham inhibited the photochemical production of ATP.

If photosynthetic inhibitions are responsible for reductions of growth, an external carbohydrate supply may prove protective, because the ATP required for growth could be obtained through the mitochondrial oxidative phosphorylation system. We found that when glucose was included in diuron-treated cultures, both growth and ATP levels returned to control levels, whereas inclusion of glucose in chlorpropham-treated cultures was essentially without

46

effect (St. John, 1971). Therefore, chlorpropham also interfered with the oxidative production of ATP, but diuron did not. Furthermore, chlorpropham had a stronger effect on ATP level than on growth for diuron, the inhibitions of growth, and ATP levels essentially paralleled each other. Thus the metabolism of diuron-treated cultures was still geared to the production of an approximately constant level of ATP, and ATP synthesis was probably not growth-limiting. Chlorpropham-treated cultures, however, did not maintain ATP in balance with growth, and ATP may have been limiting.

Our data taken collectively indicate that the luciferin-luciferase system can be adapted to studies on herbicide effects on phosphorylation processes in vivo, that effects on photophosphorylation and oxidative phosphorylation can be separated, and that herbicide effects on ATP levels can be related to growth control.

Since the physiological functions and biochemical processes required for plant growth and development are driven by energy derived from ATP, it follows that herbicides that inhibit ATP production may control growth indirectly by limiting ATP requiring biosynthetic systems. Ribonucleic acid (RNA) and protein synthesis are essential processes for plant growth and development, and, at least in *Escherichia coli*, these processes can account for up to 90 percent of the ATP expended for biosynthetic processes. When data (Moreland et al., 1969) on 14 of 22 herbicides that inhibited RNA and protein synthesis were examined, some of the strongest inhibitors of these biosynthetic reactions had been reported by various investigators to inhibit oxidative phosphorylation in isolated mitochondria. Thus, the effects of these herbicides on RNA and protein synthesis might be attributed to interference with ATP production. Gruenhagen and Moreland (1971) studied the effects of 22 herbicides on in vivo levels of ATP, orotic acid incorporation into RNA, and leucine incorporation into protein in soybean hypocotyls, using the luciferin-luciferase system to measure ATP. As expected, no herbicide was found to reduce tissue ATP levels and not inhibit RNA and protein synthesis. The correlations established between tissue ATP levels and inhibitions of RNA and protein synthesis suggested that interference with the production of energy, required to drive biosynthetic reactions, could be the mechanism through which these herbicides act.

However, herbicides could also produce phytotoxicity by interfering directly with steps in biosynthetic reaction sequences. The luciferin-luciferase enzyme system has exposed the mechanism by which such interferences could occur, and again ATP has been involved. The enzyme systems responsible for firefly light emission, amino acid activation, and fatty acid activation each form an appropriate enzyme-adenylate complex from ATP in the presence of Mg^{++} (McElroy et al., 1967). The luciferyl-adenylate complex reacts with oxygen, emitting light and releasing AMP; the amino acid-adenylate complex reacts with tRNA, forming an amino-acid-tRNA complex plus AMP: and the fatty-acid-adenylate complex reacts with CoASH, forming fatty-acid-CoA and AMP (McElroy et al., 1967).

Kinetic analyses of herbicide inhibitions of the luciferin-luciferase enzyme system in vitro done in our laboratory (unpublished data) and by Rusness and Still (1974) indicated that herbicides may be linear noncompetitive inhibitors in respect to ATP, linear competitive inhibitors in respect to D-luciferin, or linear competitive inhibitors in respect to both ATP and D-luciferin.

If the firefly luciferase system can serve as a model system to be compared to amino acid or fatty acid activation, then a second mechanism of herbicide action involving ATP can be postulated. Certain herbicides may control growth by preventing ATP use by blocking the formation of enzyme-adenylate complexes required in biosynthetic reaction sequences.

In conclusion, the luciferin-luciferase enzyme system for determination of ATP is valuable for studies on the mechanisms of herbicide action. Studies using this system have shown that certain herbicides may act by interfering with ATP production or by blocking ATP use, or possibly by both mechanisms.

REFERENCES

Gruenhagen, R. D. and D. E. Moreland, "Effects of Herbicides on ATP Levels in Excised Soybean Hypocotyls," *Weed Sci.*, **19**, 1971, p. 319-323.

McElroy, W. P., M. DeLuca, and J. Travis, "Molecular Uniformity in Biological Catalyses," *Science*, **157**, 1967, p. 150.

Moreland, D. E., S. S. Malhotra, R. D. Gruenhagen, and E. H. Shokraii, "Effects of Herbicides on RNA and Protein Synthesis," *Weed Sci.*, **17**, 1969, p. 556-562.

Rusness, D. G. and G. G. Still, "Firefly Luciferase Inhibition by Isopropyl-3-chlorocarbanilate and Isopropyl-3-chloro-hydroxycarbanilate Analogues," *Pest. Biochem. Physiol.*, **4**, 1974, p. 109-119.

St. John, J. B., "Determination of ATP in *Chlorella* with the Luciferin-luciferase Enzyme System," *Anal. Biochem.*, **37**, 1970, p. 409-416.

St. John, J. B., "Comparative Effects of Diuron and Chlorpropham on ATP Levels in *Chlorella*," *Weed Sci.*, **19**, 1971, p. 274-276.

BIOLUMINESCENCE FOR DETERMINING ENERGY STATE OF PLANTS

Te May Ching
Crop Science Department
Oregon State University
Corvallis, Oregon

The bioluminescence produced by the luciferin-luciferase system is a very sensitive assay for ATP content in extracts of plant materials. With two additional enzymes, pyruvate kinase (E.C. 2.7.1.40) and adenylate kinase (E.C. 2.7.4.3), and an excess of phosphoenolpyruvate, the ADP and AMP in tissue extracts can be quantitatively converted to ATP and then assayed with the luciferase system (Ching and Ching, 1972). This simple, rapid, and sensitive procedure has been used for determining the energy state of plant materials for metabolic studies (Ching and Ching, 1972; Ching et al., 1974) under different environmental stresses,* in cultivars of varied genetic capabilities (Ching and Kronstad, 1972), and in seeds of different viability and vigor (Ching, 1973; Ching and Danielson, 1972). The experimental results from our laboratory and others indicate that the bioluminescence test is a very useful tool for (a) predicting viability of seeds and pollens in a time period of one hour instead of the usual days or weeks required by growth tests; (b) screening cultivars for high growth potential and productivity without long-term and labor-demanding field tests; (c) detecting the incipient damage of environmental pollutants and stresses so that reparative measures can be implemented; and (d) discerning the role of ATP concentration, total content of adenosine phosphates, ATP/ADP ratio, and energy charge [EC, EC = ([ATP] + 1/2 [ADP])/([ATP] + [ADP] + [AMP]) – (7)] in metabolic control of enzymes, organelles, tissues, and organisms. All these possibilities will be discussed in the following sections in more detail and particular applicational requirements will be mentioned.

ATP TEST FOR SEED AND POLLEN VIABILITY AND VIGOR

ATP is needed for biosynthesis of nucleotides, nucleic acids, proteins, lipids, amino acids, and carbohydrates (Henderson and Paterson, 1973). ATP also modifies existing enzymes in cells and tissues, thus it activates, inactivates,

---

* Barlow, E. W. R., "Physiological Effects of Water Stress on Young Corn Plants," Ph.D. thesis, 1974, Oregon State University.

or amplifies the enzyme activity, depending on the condition (Holzer and Duntze, 1972). In seeds as well as pollens, ATP is usually low and often limiting for the early metabolic events of germination (Ching, 1972; Obendorf and Marcus, 1974). If more ATP can be synthesized, the metabolic wheel will be turned faster, which eventually leads to more growth. Figure 1 shows such correlation in rape seeds. Further indication of high ATP and energy charge associated with high synthetic ability of ribonucleic acid (RNA), protein, and lipids in wheat embryos or seedlings are summarized in table 1. In pollen of douglas fir (*Pseudotsuga menziesii* Franco), the ATP contents were found to be 50, 136, 209, 292, 456, and 562 n moles per one gram of fresh weight for lots having 0, 9, 16, 30, 57, and 91 percent germination, respectively. Again, the data indicate a high correlation of ATP content and germinability (unpublished data from our laboratory). All these data show that among the seed or pollen lots of one crop or cultivar, ATP content appears to be a good biochemical index of vigor and viability. In order to be of practical use for seed testing or cross pollination work, a standard curve depicting the relationship of ATP content and growth potential or germinability should be established for each cultivar of a crop. At present, several

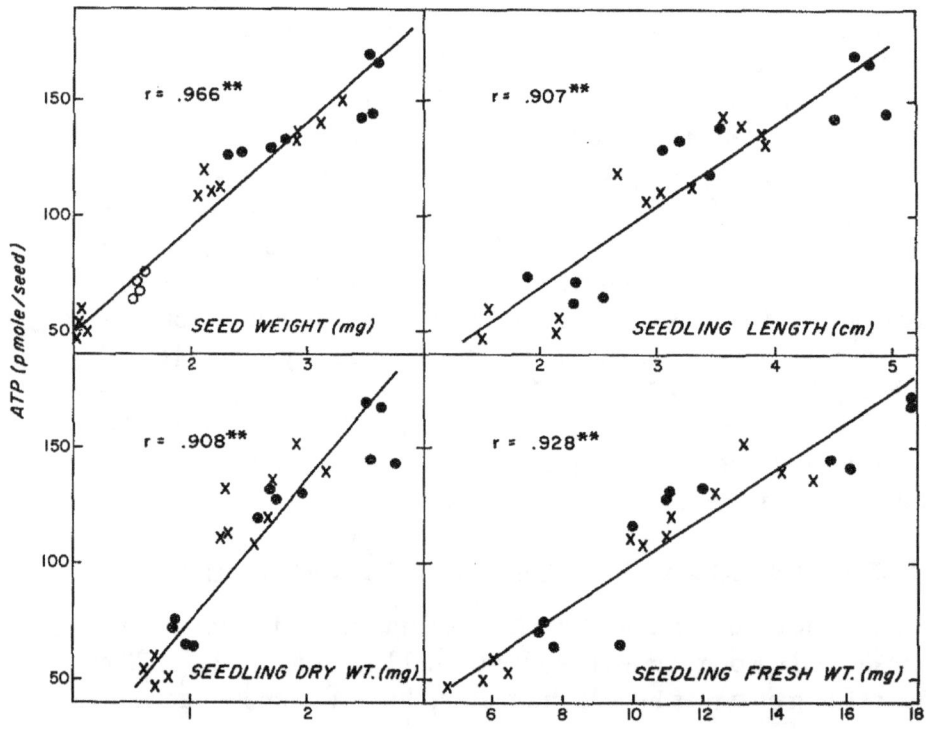

Figure 1. Correlation of ATP content in 4-hour imbibed seeds of rape (*Brassica napus* L.) and seed weight, 4-day seedlings length, 4-day seedling fresh and dry weight. X, 0 = different lots; r = correlation coefficient; ** = significant at 1 percent level (Ching and Ching, 1973).

Table 1

Contents of ATP and Total Adenosine Phosphates (TAP), Energy Charge
(EC), Protein, RNA, and Lipid Synthesizing Ability and Growth Rate
of the Embryo or Seedling of Wheat Cultivars, Hyslop (C.I. 14565)
and Yamhill (C.I. 14563) (Ching, 1972)

	Hyslop	Yamhill
Embryo dry weight, mg	1.19 ± 0.13	1.26 ± 0.14
4-hours embryo: ATP, pmole	232 ± 29	290 ± 35
TAP, pmole	504 ± 56	620 ± 41
EC	0.71	0.72
2-day seedling: ATP, pmole	2985 ± 315	4800 ± 360
TAP, pmole	4819 ± 501	6656 ± 702
EC	0.81	0.86
UL–^{14}C-amino acid incorporation		
(cpm/5 seedlings): 30 hours old	1526 ± 184	2138 ± 329
44 hours old	4701 ± 301	7626 ± 513
5–^{3}H-uridine incorporation		
(cpm/5 seedlings): 20 hours old	1060 ± 172	1259 ± 140
36 hours old	2438 ± 313	5897 ± 647
UL–^{14}C-glucose incorporation		
into lipids (cpm/5 seedlings):		
24 hours old	1415 ± 152	1760 ± 154
48 hours old	3229 ± 289	6242 ± 344
7-day seedlings: fresh weight mg	157 ± 12	201 ± 15
dry weight mg	18 ± 1	21 ± 1

standard curves have been established for cultivars of lettuce seeds (Ching and
Danielson, 1972). More standard curves for different crops are therefore
needed for this test.

PREDICTOR FOR HIGH GROWTH POTENTIAL AND PRODUCTIVITY IN NEW CROSSES AND SELECTIONS OF BREEDING MATERIALS

The data in table 1 indicate that between cultivars of the same crop, differences exist in synthetic ability, and that a positive correlation of high ATP
content, total adenosine phosphates, and energy charge with high synthetic
ability is clearly demonstrated. A time curve showing the fluctuation of
ATP, ADP, and AMP pools in the wheat embryo or seedling further depicts

the dynamic aspect of energy metabolism in germinating seeds (figure 2). It shows clearly that three oscillations of synthesis and utilization of adenosine phosphates occurred in the wheat embryo and seedlings during the first 48 hours of germination. Each dip on the curve coincides with major metabolic and morphological events. The first one associates with protein and RNA synthesis (Ching, 1972; Mazus' and Buchowicz, 1973); the second one follows DNA synthesis (Ching and Kronstad, 1972) and root emergence; and the third dip relates with the shoot emergence and the myriad of synthetic activities that followed. One trend shows clearly that the cultivar Yamhill with higher growth potential had higher ATP utilization and synthesis ability than Hyslop (table 1). Furthermore, the rate of synthesis exceed utilization much more in Yamhill, resulting in a larger pool of ATP and ADP and a higher energy charge in Yamhill seedling. The larger pool and high energy charge in turn facilitates more biosynthesis and faster growth which eventually could lead to more productivity under optimum conditions. This speculation was born out by a cooperative study with Dr. R. G. McDaniel at the University of Arizona on nine newly developed Mexican wheat cultivars. In the study we found that the efficiency of oxidative phosphorylation of isolated mitochondria and the adenylate energy charge of germinating seedlings are positively significantly correlated with grain yield (correlation coefficients of +0.729* and +0.687*, respectively). Therefore, we have concluded that both biochemical parameters appear to be useful tools for plant breeders to

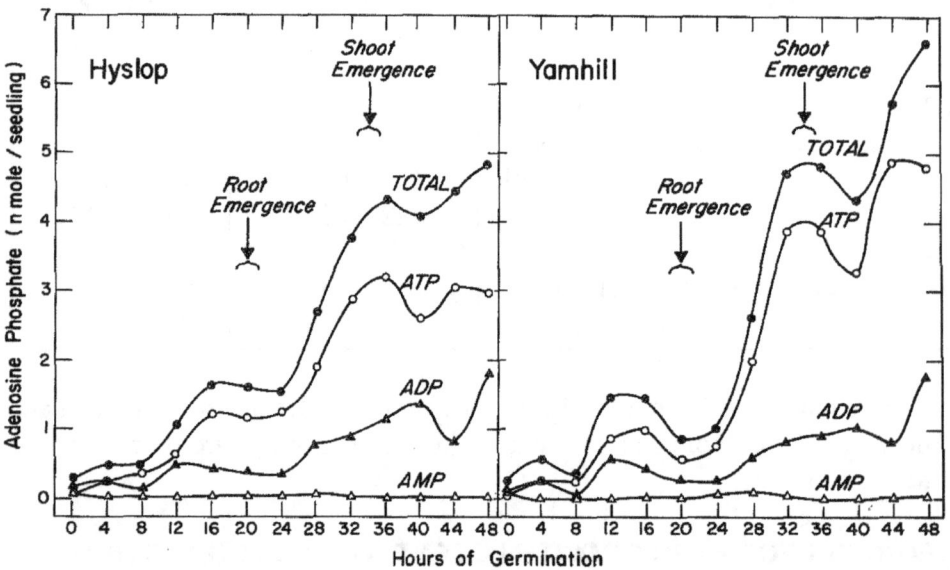

Figure 2. Changes in the content of ATP, ADP, AMP, and total adenosine phosphates in the embryo and the seedling of Hyslop and Yamhill wheat seed germinated at 293 K (20° C) in the dark (Ching and Kronstad, 1972).

---

* Significantly correlated at the 5 percent level.

screen for high producers at an early stage of a breeding program to reduce the time- and money-consuming field trials. More information is needed in the application of this concept in actual programs.

INDICATOR FOR ENVIRONMENTAL QUALITY AND STRESSES

Salts in soil reduced the growth and the ATP content in pea roots (Hasson-Porath and Paljakoff-Mayber, 1971). Cold temperature decreased ATP in cotton seedlings and germinating douglas fir seeds (Stewart and Guinn, 1969; Ching and Ching, 1973). Herbicides (for example, chloropropham) lowered the ATP level and inhibited RNA synthesis in soybean hypocotyls (Gruenhagen and Moreland, 1971), and atmospheric pollutants (for example, ozone) reduced ATP content and energy charge in douglas fir seedlings.* Water stress or drought conditions caused a reduction of ATP content in corn leaves† and germinating douglas fir seeds (Ching and Ching, 1973). From all these reports, one may develop a procedure with which to monitor the environmental influences and estimate the degree of damage or benefit so that reparative measures can be implemented.

STUDIES ON METABOLIC REGULATION

The regulation of carbohydrate metabolism by adenine nucleotides was known in 1964. Subsequently, Atkinson developed the concept of energy charge as an overall measure of the energy state of cells (Atkinson, 1969). Based on experimental data, Atkinson and his colleagues observed that when the energy charge is greater than 0.5, ATP-utilizing systems increase their activities, and, at an energy charge lower than 0.5, ATP-regenerating sequences become dominant. Therefore, the energy charge could modulate biosynthesis in cells and tissues.

In plant materials (Davies, 1973), information regarding the regulation by adenine nucleotides are very scanty. ATP and citrate inhibited phosphofructo-kinase in carrot, corn scutellum, and peas, whereas ADP and AMP were relatively ineffective in modulating the response to ATP. The plant isocitrate dehydrogenase responded neither to individual adenosine phosphates nor to energy charge, but the decarboxylation of alpha oxoglutarate was stimulated by AMP in the mitochondria of pea and cauliflower. Recently, the *first positive evidence of energy charge controlling the activity of 3-phosphoglyceric acid kinase from pea leaves* was reported (Pacold and Anderson, 1973). The ribulose-5-phosphate kinase of the same tissue, however, was

---

* Tingey, D., National Ecol. Res. Lab., U.S. Environmental Protection Agency, Corvallis, Oregon, private communication.

† Barlow, E. W. R., "Physiological Effects of Water Stress on Young Corn Plants," Ph.D. thesis, 1974, Oregon State University.

not controlled by the energy charge. The difference is difficult to resolve and awaits future experimentation.

The data in table 1 indeed indicate a modulation of synthesis by energy supply and energy charge. At different times of seed germination and maturation, such positive correlations do not exist, however, particularly at the peak of cell division (Ching and Ching, 1972) and the peak of synthesis of nucleotides and nucleic acids (Ching et al., 1975) (figure 3). Apparently a temporal control of development overrides the energy charge regulation or simply because the end product inhibition does not occur in the biosynthetic pathways of nucleotides (Henderson and Paterson, 1973). In tissues with stabilized cell numbers, such as the gametophyte of ponderosa pine seeds, an increased ATP content, total adenosine phosphates, and energy charge preceded the biogenesis of enzymes and organelles and then all reduced in concert after the function of these enzymes and organelles was fulfilled (Ching, 1970) (figures 4,5).

It is clearly shown from this discussion that many regulatory mechanisms are operative in the complex systems of eukaryotes, but few are absolutely established yet. The bioluminescence assay of adenosine phosphates will definitely facilitate the exploration of these mechanisms.

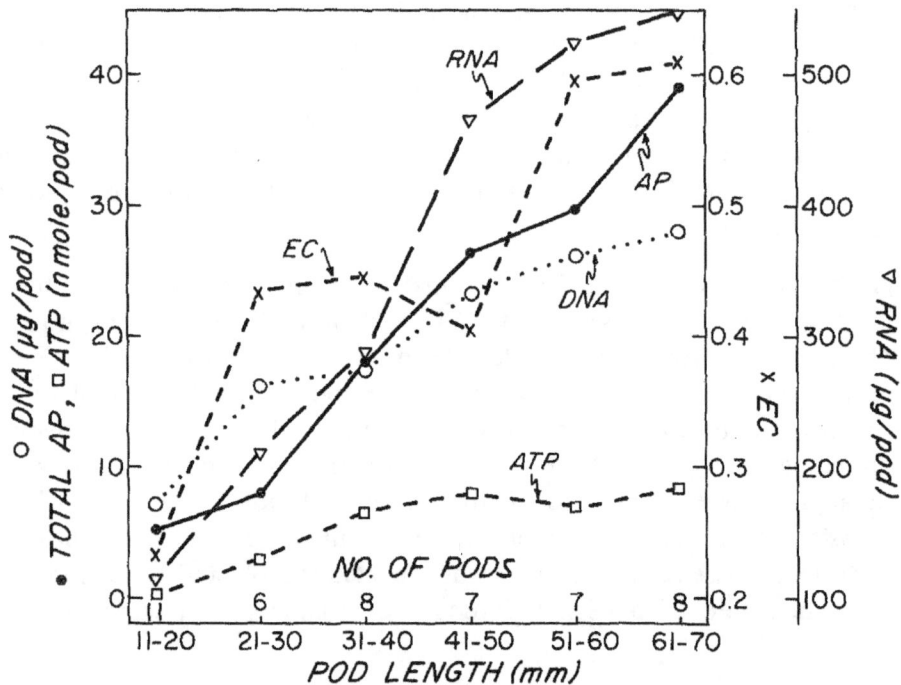

Figure 3. Contents of ATP, total adenosine phosphates (AP), DNA and RNA, and energy charge in whole 8-day-old pods of rape (Ching, et al., 1974).

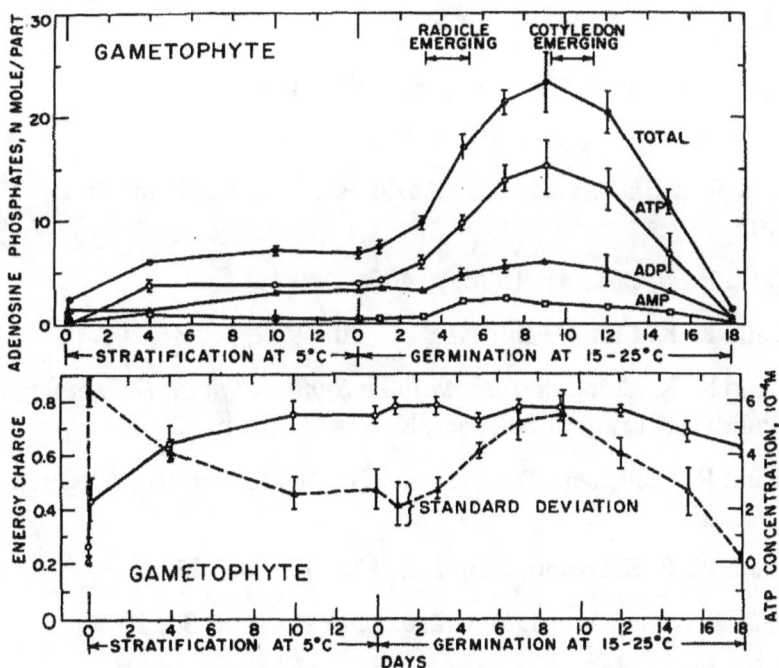

Figure 4. Changes in the content of ATP, ADP, AMP, and total adenosine phosphates (upper) and changes in adenylate energy charge (—) and ATP concentration (---) (lower) in the gametophyte of germinating ponderosa pine seeds (Ching and Ching, 1972).

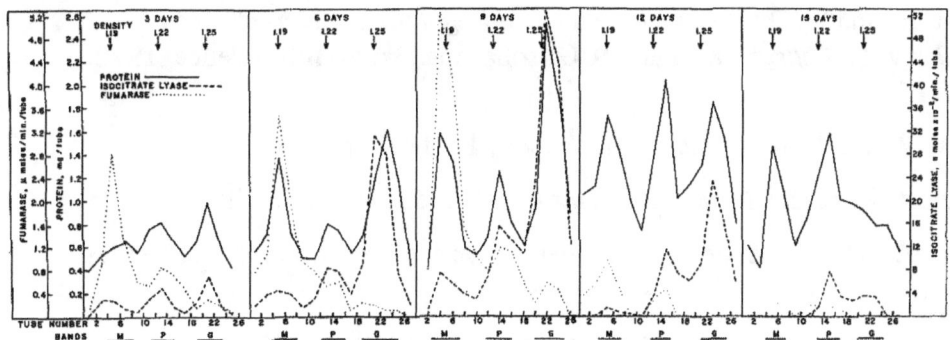

Figure 5. Changes in protein content and activity of fumarase and isocitrate lyase with germination days in fractions isolated by sucrose density gradient centrifugation of 10,000-g pellet from 250 gametophytes of germinating ponderosa pine seeds. M = mitochondria; P = mixture; G = glyoxysomes (Ching, 1970).

REFERENCES

Atkinson, D. E., *Annu. Rev. Microbiol.*, **23**, 1969, p. 47.

Ching, T. M., *Plant Physiol.*, **46**, 1970, p. 475.

Ching, T. M., in *Seed Biology*, **II**, T. T. Kozlowski, ed., Academic Press, 1972, p. 103.

Ching, T. M., *Plant Physiol.*, **51**, 1973, p. 400.

Ching, T. M. and K. K. Ching, *Plant Physiol.*, **50**, 1972, p. 536.

Ching, T. M. and K. K. Ching, in *International Symposium on Dormancy in Trees*, Polish Academy of Sciences, Kornik, 1973, p. 13.

Ching, T. M. and R. Danielson, *Proc. Assoc. Off. Seed Analysts.*, **62**, 1972, p. 116.

Ching, T. M. and W. E. Kronstad, *Crop Sci.*, **12**, 1972, p. 785.

Ching, T. M., D. L. Stamp, and J. M. Crane, *Plant Physiol.*, **54**, 1974.

Davies, D. D., in *Biosynthesis and its Control in Plants*, B. V. Milborrow, ed., Academic Press, 1973, p. 2.

Gruenhagen, R. D. and D. E. Moreland, *Weed Sci.*, **19**, 1971, p. 319.

Hasson-Porath, E. and A. Paljakoff-Mayber, *Plant Physiol.*, **47**, 1971, p. 109.

Henderson, J. F. and A. R. P. Paterson, *Nucleotide Metabolism*, Academic Press, 1973, p. 54.

Holzer, H. and W. Duntze, in *Biochemical Regulatory Mechanisms in Eukaryotic Cells*, E. Kun and S. Grisolia, eds., Wiley-Interscience, 1972, p. 115.

Mazus', B. and J. Buchowicz, *Phytochem.*, **12**, 1973, p. 271.

Obendorf, R. L. and A. Marcus, *Plant Physiol.*, **53**, 1974, p. 779.

Pacold, I. and L. E. Anderson, *Biochem. Biophys. Res. Comm.*, **51**, 1973, p. 139.

Stewart, J. M. and G. Guinn, *Plant Physiol.*, **44**, 1969, p. 605.

ULTRASENSITIVE BIOLUMINESCENT DETERMINATIONS OF ADENOSINE TRIPHOSPHATE (ATP) FOR INVESTIGATING THE ENERGETICS OF HOST-GROWN MICROBES

J.H. Hanks and A.M. Dhople
*Johns Hopkins-Leonard Wood Memorial Leprosy
Research Laboratory
Baltimore, Maryland*

In all living things, biologically useful energy is captured in adenosine triphosphate (ATP). Our interest in ATP arose from the need for a biologically significant biochemical tool applicable to host-dependent microbes. The theoretical basis for regarding ATP as a key compound appears to be sound.

- The concentration of ATP pools within any species under defined different circumstances is controlled by the net balance between rates of generating energy and rates of biosynthesis.

- Minimal levels of ATP suffice for energy of maintenance; slightly higher levels stimulate minimal rates of growth.

The practical question was whether the desired sensitivity could be obtained for metabolic investigations of host-grown microbes, which have been handicapped by the great number of cells required and by the fact that "host-grown" species have not been stabilized or activated in vitro.

McElroy and associates were the first to demonstrate the role of ATP in illuminating the tails of fireflies and developed methods whereby the bioluminescence can assay ATP levels. The almost miraculous sensitivity of this principle has been refined by Chappelle and associates at the Goddard Space Flight Center. In view of the possibility of achieving a sensitivity applicable to routinely obtainable numbers of *Mycobacterium leprae* cells, we undertook to confirm the stability of the reagents and to define the optimal concentrations of each of the reactants.

Luciferase enzyme was prepared and purified by the method of Chappelle using Sephadex G-100. The lyophilized enzyme in the presence of luciferin and magnesium, when stored at 197 K (-76°C), is stable for over 30 months. When rehydrated and held at 273.6 to 277 K (0.5 to 4°C), 96 percent of the original activity is retained after 7 days and 93 percent is retained after 11 days. One unit of enzyme is defined as the amount of enzyme per 0.3-ml acceptor system containing 100-μg/ml luciferin and 0.005 M Mg^{++}, that elevates the peak of ATP response curves 50 units (1 cm (2.5 in.)) due to 10 pg ATP. We denote this system as R-system (routine) and use it for all routine purposes. Our H-system (high sensitivity) contains 3 units of enzyme in the presence of 300 μg/ml luciferin and 0.0075 M Mg^{++} and gives an elevation of 150 units with 10 pg of ATP. The character of the interactions between 10 pg of ATP and acceptor systems containing different concentrations of luciferase, luciferin, and magnesium were investigated systematically.

Table 1 shows the method and the results of titrating optimal balances between the three interacting components of ATP acceptor systems and demonstrates the effects of luciferin concentration. The data in the columns for 1 and 3 units of enzyme clearly define the optimal concentrations of reactants for the R- and H-systems.

Table 1
Interactions and Optimal Concentrations of
Luciferase, Magnesium, and Luciferin*

Enzyme units:	1				2			3			4		
Mg (mM)	2.5	5.0	7.5	10.0	5.0	7.5	10.0	5.0	7.5	10.0	5.0	7.5	10.0
Luciferin (μg/ml)													
50	23	46	29										
(65)†			(33)										
100	25	(50)	37	27	68	75	66						
200		47	35		89	109	86	115	122	108	110	121	105
300		45	41		118	125	100	145	(150)	125	143	150	134
400	40	44	46	39	116	122	104	139	143	128	129	145	130

* All reactants = concentrations during bioluminescence.
  All data = units of response/10 pg ATP.
  Values circled = optima for the R- and H-systems.

† Sensitivity of the system of Chappelle and Levin (*Biochem. Med.*, **2**, 1968, p. 41-52) in the presence of 1 unit enzyme.
  The R-system uses only 92 percent as much enzyme, but is 1.5 times more sensitive.

In figure 1, panel A illustrates interdependencies between enzyme and luciferin concentrations in presence of optimal Mg. The effects of increasing enzyme concentration in the two acceptor systems differed markedly. When the luciferin and Mg were balanced optimally for one enzyme unit, they were inadequate for higher concentrations of enzyme. With the concentrations

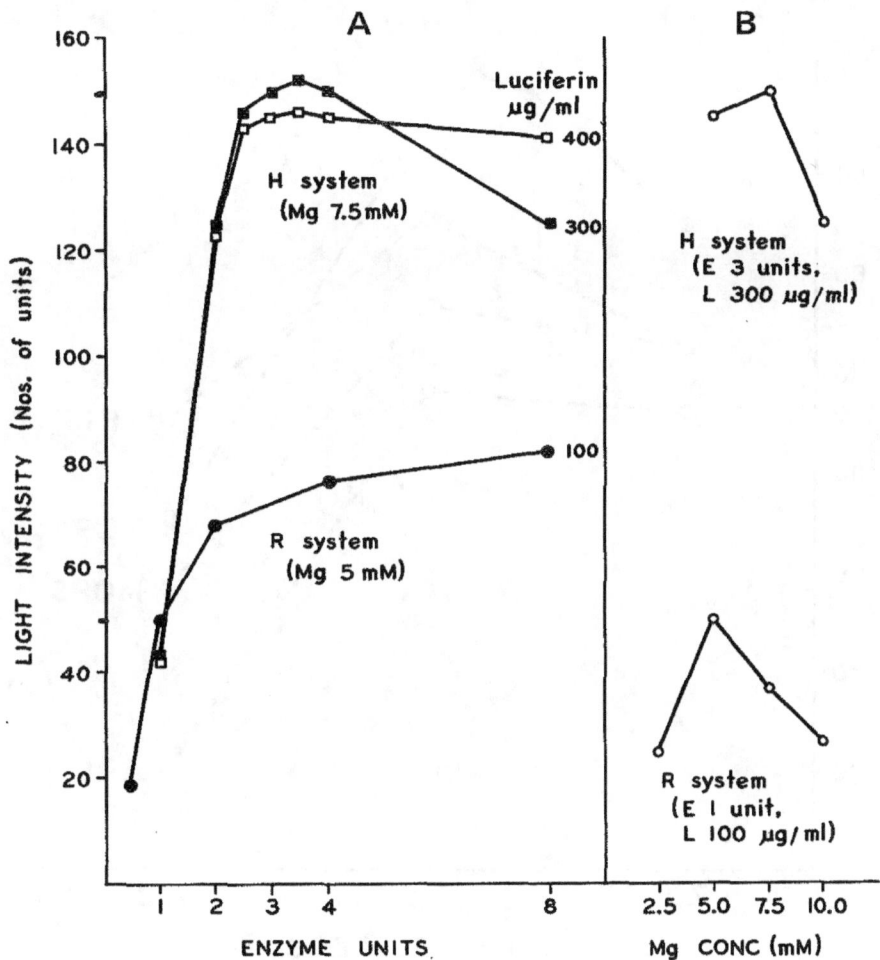

Figure 1. Panel A—the effect of enzyme concentration. Panel B—the effect of Mg concentration.

of luciferin (300 μg/ml) and Mg (0.0075 M) appropriate to the H-system, an increase from 1 to 2.5 units of enzyme caused a linear increase in sensitivity. Three and 3.5 enzyme units produced further, limited gains prior to the onset of inhibitions, which suggest that enzyme purity is a major limiter of sensitivity.

The data on panel B in figure 1 are restricted to optimal concentrations of enzyme and luciferin for the R- and H-systems and illustrate the differing optima with respect to Mg concentration. There were always losses of sensitivity when the designated optimal concentrations were altered.

The optimal ionic strength was confirmed to be 0.05 M for the four buffers tested. Figure 2 compares the luminescence in the presence of four buffers at pH 7.0, 7.5, and 8.0. The fact that a pH higher than 7.5 was always disadvantageous suggested that the optimal pH might be below 7.5. As

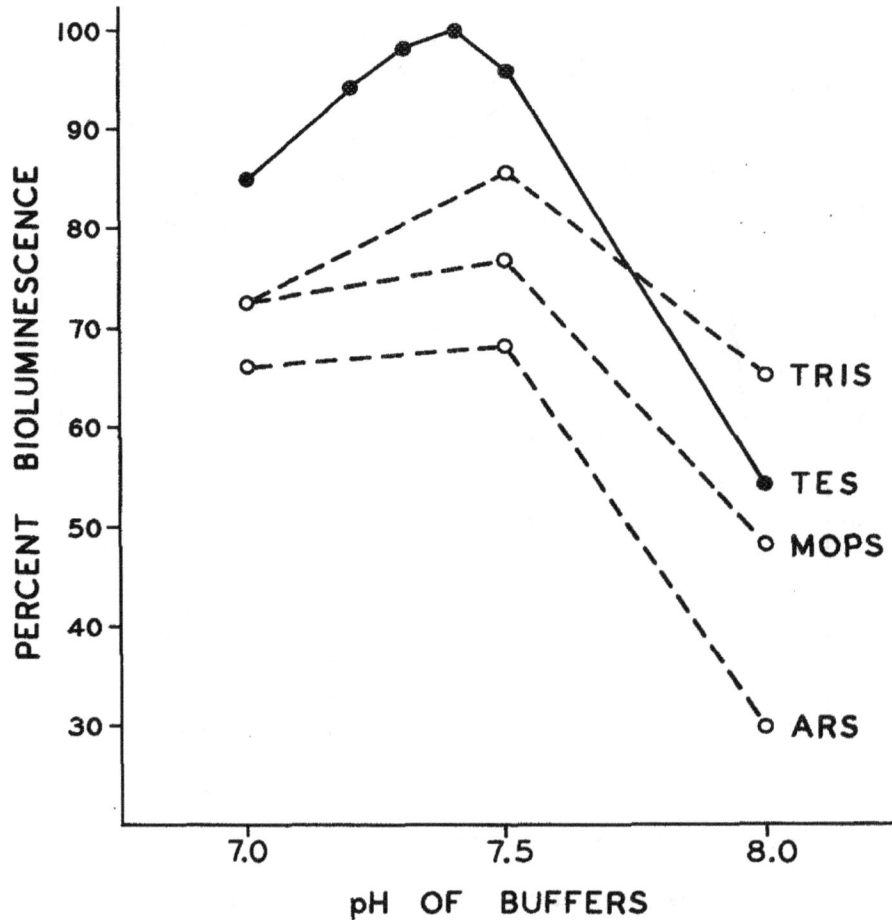

Figure 2. Comparison of luminescence in the presence of TRIS, TES, MOPS, and ARS buffers.

shown, pH 7.4 was optimal for the TES (N-tris [Hydroxymethyl] methyl-2-aminoethane sulfonic acid) buffer adopted. This pH provides the best margin of safety against errors in pH adjustments.

The features of the R- and H-systems as finally adopted are summarized in table 2. The two major points are that:

● The optima for buffer and pH are identical, and

● The optimal concentrations of enzyme, magnesium, and luciferin depend upon the sensitivity desired.

Before explaining the reference method for extracting ATP, I will present the procedures required to investigate the energetics of unwashed, host-grown microbes. In such cases, one must faithfully eliminate host ATP before releasing the bacterial ATP (see procedure 1). The first step in procedure 1 eliminates 99 percent of any soluble, nonbacterial ATP in the

Table 2
Optimal Concentrations of Bioluminescence Regulators

Acceptor:	*R-system	†H-system
Luciferase	1 unit	3 units
Luciferin	100 μg/ml	300 μg/ml
MgSO$_4$	0.005 M	0.0075 M
Buffer	TES, 0.05 M	TES, 0.05 M
pH	7.4	7.4

* R = Routine, for cultivable organisms.
† H = High sensitivity, for host-dependent organisms.
All reactants: concentrations during bioluminescence.

Procedure 1
Elimination of Host ATP and Extraction of Microbial ATP

Purpose of steps	Procedure	Dilution of soluble ATP
1. Remove soluble ATP	0.1 ml bacteria suspended in 10-mm or 13-mm glass tube +0.9 ml TES buffer. Centrifuge, cold, 20 min at 3,000 g; remove 99% supernatant with a Manostate Accropet. Resuspend sediment in 0.1 ml TES buffer.	100 X
2. Extract ATP	Add 0.03 ml chloroform (23% v/v), vortex 10 s. 331 K (98°C) water bath: (a) heat n min*; (b) apply 750-mm vacuum with shaking for 1 min to yield dry samples.	
3. Assay of ATP	Rehydrate in 0.4 ml TES buffer. Assay immediately by injecting 0.1 ml into 0.3 ml acceptor system.	4 X
Total dilution (safety) factor:		400 X

* *E. coli* = 1 min; *M. phlei* = 4 min; *M. lepraemurium* = 9 min.

61

sample. The second step extracts the bacterial ATP and produces a dried sample. After the third step, the bacterial ATP is at the volume required for triplicate assays. Meanwhile, the mammalian ATP has been diluted 400 times.

Each step in the foregoing methods was developed and validated by several criteria:

- By means of known amounts of ATP added to bacterial suspensions to represent soluble ATP;

- By using single-cell suspensions of *Mycobacterium phlei* and performing microscopic counts, plate counts, and ATP determinations; and

- By using both active and extracted *Mycobacterium lepraemurium* cells throughout the pertinent features of each promising method.

The classical extractors of ATP—PCA and n-butanol—were found incapable of extracting total microbial ATP, even from *E. coli*. These agents released only 70 to 85 percent of the ATP from a *saprophytic mycobacterium, M. phlei,* and some 50 to 70 percent of that in the pathogenic *M. lepraemurium*. Neutralization of PCA diluted the extracts and spoiled sensitivity. N-butanol quenched bioluminescence, even when diluted to 0.1 percent.

Table 3 summarizes the percentages of ATP pools released from representative microbes by heat and chloroform.

Previous investigations had uncovered only two agents—heat and chloroform—which promptly opened the *M. Lepraemurium* cells to dye penetration. Present work evolved a convenient reference method having minimal effects on reference standard ATP, giving maximal yields of dried ATP and no ATP upon reextraction of cell residues. The merits of chloroform are:

- It does not degrade reference standard ATP;

- It disrupts enormous clumps of mycobacteria instantaneously, thus exposing the surfaces of individual cells; it disrupts cell membranes; and

- Because of its low boiling point, it is readily removed by heat and cannot quench bioluminescence.

We have shown that heat at 331 K (98°C) degrades ATP so slowly that 98 percent of the reference standard ATP is assayable after 10 minutes. After sequential application of chloroform and heat for 10 minutes, 96 percent of the reference standard ATP was assayable.

The significant points in table 3 are:

- Items 1 and 2—neither heat nor chloroform alone suffice, even for *E. coli*. The other species were progressively more resistant, and

Table 3
Extraction of ATP by Heat and Chloroform
Basis of Reference Methods

Procedures	331 K (98°C)* (min)	% ATP pool released from: †		
		E. coli	M. phlei	M. leprae-murium
1. Heat only:	2	52	45	31
	5	78	69	54
	10	90	88	56
2. Chloroform only: ‡	–	94	93	62
3. Heat first, then chloroform: §	5	100	101	79
	10	–	–	100
4. Chloroform first, then heat: ‖	2	100	96	70
	5	–	100	86
	10	–	–	100

* Data = heating prior to vacuum drying.
† Calculated as shown previously (Dhople and Hanks, *Applied. Microb.*, **26**, 1973, p. 399-403).
‡ Chloroform = 0.03 ml added to 0.1 ml samples = 23% v/v; vortexed 10 s.
§ Reference procedure 3. Terminal heating = 1 min to warm the sample plus 1 min of vacuum drying.
‖ Reference procedure 4. Terminal heating = 1 min to vacuum dry.

chloroform was a more effective disruptor of membrane-wall complexes than heat.

● Item 4—Applying chloroform as the first step in the sequential use of the two agents was the most convenient method. The optimal heating period was 2 minutes for *E. coli*, 5 minutes for *M. phlei*, and 10 minutes for *M. lepraemurium*.

The overall gains in sensitivity for quantitation microbial ATP can be summarized as follows:

1. The R-system is 1.5 times more sensitive than that of Chappelle and Levin. The H-system is 3 times more sensitive than the R-system, a total gain of 4.5 times.

2. Because of a novel method of extracting total microbial ATP, the number of cells required per sample has been decreased to ±15 times.

3. Thus, 1.4 percent of the cell numbers originally required suffice for analyses of ATP.

Three years of working experience have enabled us to define the limits of sensitivity and the minimal concentrations of ATP that can be quantitated at present. These are shown in table 4.

Table 4
Quantitations of ATP

A. Limits of Sensitivity			B. Reliability
pg ATP/ assay	% Reference standard ATP assayed		
	†R-system Units %	†H-system Units %	
3 1 0.3 0.1	14 93 3 60 D*	43 96 14 93 3 67 D	1. The standard deviations of duplicate determinations of ATP are equivalent to those for triplicate plate and microscopic counts. 2. Average SDs, expressed as percentages of the observed values = ± 4%.

*   D = detected, not quantitated.
†   R = routine; H = high sensitivity.

- Sensitivity. The efficiency of demonstrating ATP at less than 7 pg per assay falls off on a parabolic curve. This table sets the limits of the R-system at about 1 pg and that of the H-system at 0.3 pg. Correction factors compensate for the declining efficiency and yield data based on reference standard ATP.

- Reproducibility. The two goals were (a) to turn out duplicate assays having the same precision as triplicates and (b) to obtain a reproducibility of ± 2.5 percent. This was accomplished by means of mental arithmetic. When paired determinations agreed within 5 percent (that is, ± 2.5 percent of the mean), the results were accepted. When one value was more than 5 percent less than the higher value, a further (triplicate) assay was made. A review of accumulated data showed that triplicate assay had been required in only 10 percent of the total samples. Thus, 2.2 units of work yielded higher reproducibility than can be obtained from triplicates which are not monitored mentally while the work is in progress.

The estimate of ± 4 percent in table 4 was caused by the fact that the base data included determinations on bacterial suspensions prepared on different dates.

In conclusion, we have (1) redefined the optimal concentrations of the five reactants in the bioluminescent system, (2) devised a novel method of eliminating host ATP while extracting and drying total microbial ATP, and (3) now require only 1.4 percent of the number of cells originally needed.

The R-system for routine work is more efficient than the previous systems, while the H-system quantitates as little as 0.3 pg ATP. The usual triplicates can be replaced by 2.2 assays per sample, with a gain in reproducibility.

ENERGETICS OF THE HOST-DEPENDENT
MYCOBACTERIUM LEPRAEMURIUM
DURING TRANSITION TO A CAPACITY
FOR EXTRACELLULAR GROWTH

A. M. Dhople and J. H. Hanks
Johns Hopkins-Leonard Wood Memorial Leprosy Research Laboratory
Baltimore, Maryland

The possibility of cultivating so-called host-dependent microbes (HDM) is fascinating. We now focus on *Mycobacterium lepraemurium*, the previously "obligate intracellular agent" of rat leprosy. Since this organism, until recently, had been noncultivable for 70 years, it is a model from which one can derive principles applicable to the urgently needed cultivation of *M. lepraemurium*. Metabolic or biochemical monitors of the physiologic state of the experimental materia are a sine qua non because (1) such methods measure inhibitory conditions as readily as favorable ones, and (2) the lack of normal enzyme content makes host-dependent organisms particularly susceptible to inhibition.

In order to develop a biologically significant biochemical tool for quantitating the energetics and growth potential of unwashed, host-grown microbes during the progression, regression, and therapy of diseases such as leprosy, we have taken advantage of the fact that ATP can be measured in picogram amounts; that it is the source of energy for biosynthesis; and that ATP data can be interpreted either in terms of functional biomass or growth potential as follows:

- ATP pools are extremely labile, either used rapidly for biosynthesis, exchanged with related nucleotides, or in damaged cells, degraded rapidly by ATPase.

- Under constant conditions (for example, in vivo), the ATP pools within bacterial cells are controlled by the net balance between rates of generating energy and rates of biosynthesis. ATP per aliquot or per culture measures functional biomass or cell numbers.

- Minimal levels of ATP suffice for energy of maintenance; maximal levels promote maximum growth rates. Thus, ATP per bacterial cell can rank suspensions of a given species in terms of growth potential.

Because ATP data afford a means for distinguishing between genuine and spurious microscopic growth, the ATP system was utilized to investigate two reports of the microscopic growth of *Mycobacterium lepraemurium* in extracellular environments. In order to standardize conditions for incubations both in vitro and in mice, all inoculums consisted of a few million *M. lepraemurium* cells enclosed in Rightsel-Ito type diffusion chambers.

Oiwa in 1967 reported slow growth of *M. lepraemurium* when adsorbed onto siliconed slides and supplied daily with a medium containing 10 percent serum and filtered mouse brain extract. All components were prepared and handled as for mammalian cell cultures. In procedure 1 it is seen that this medium provided redundant sources of nitrogen and vitamins and practically everything required by fastidious microbes or mammalian cell cultures.

In the Rightsel system, the chambers were implanted in the mouse peritoneal cavity for 50 days before being analyzed. The results obtained in these two types of experiments are presented in figure 1.

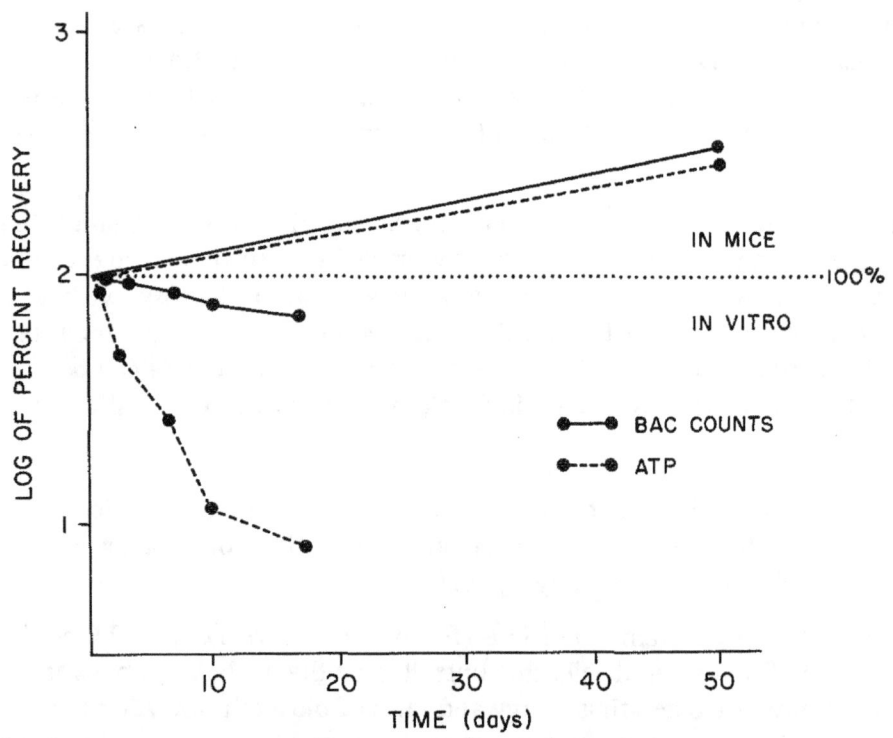

Figure 1. Bacteria counts and ATP in extracellular *M. lepraemurium* cells incubated in mice and in vitro.

In the Oiwa medium (see figure 1), irrespective of incubation system, modifications of brain extracts, or frequency of renewals, the energetics of the

Procedure 1
Preparation of Basal Medium and Brain Extract

Ingredients	Original (Oiwa)	Modified (Basal Medium)
KH_2PO_4	4.0 g	---
Na_2HPO_4, $12H_2O$	3.0	---
Sodium glutamate	1.0	---
Succinic acid	0.2	---
Magnesium sulphate	0.6	0.6 g
Fe Ammonium citrate	0.4	0.4
Sodium citrate	2.0	2.0
L-Asparagine	2.0	2.0
L-Glutamine	1.0	1.0
Casamino acids	2.0	2.0
Glucose	10.0	10.0
Sucrose	76.0	76.0
Yeast extract	5.0	5.0
Cytochrome - c	20 mg	20 mg
Ca-pantathonate	200	200
ATP	200	200
Distilled water to make	1,000 ml	900 ml
pH	6.8	6.8
Sterilization	Seitz	Seitz
Add aseptically:		
Penicillin (10,000 γ/ml)	1.0 ml	1.0 ml
Bovine serum	100 ml	100 ml

Brain Extract

Remove brains from young mice; prepare 10 percent (w/v) extract
in the above medium; centrifuge for 60 min in cold at
2,000 X g; supernate turbid.
Centrifuge the supernate for 2 hours in cold at 18,000 X g;
supernate clear.
Portion of the above brain extract pasteurized by heating at
333 K (60°C) for 30 minutes.
Keep frozen until ready to use.

M. lepraemurium cells declined steadily and uniformly. After 17 days of incubation, 72 percent of the original number of bacilli was recovered. Meanwhile, the ATP per cell had decreased to 9 percent of the original values,
leaving no possibility that genuine growth could occur.

Very different results were obtained from chambers incubated by Rightsel in the peritoneal cavities of mice. After 50 days, the bacterial biomass had increased 2.7-fold and the ATP per culture 2.5-fold. Since ATP per cell was 93 percent of the original, the Rightsel system is regarded as the first to permit extracellular growth of a so-called "obligate intracellular microbe." It follows that *M. lepraemurium* is capable of being cultivated in bacteriologic media.

In 1972, Nakamura (see procedure 2) from Japan reported that *M. lepraemurium* undergoes microscopic growth in a nonconventional system, characterized by restricted air volume and the inclusion of 1-cysteine in the medium. The autoclaved base, to be used as a control medium, is adequate for the growth of tubercle bacilli. The filtered supplements for the complete medium, NC-5, are listed in the right-hand column of procedure 2.

Procedure 2
The Nakamura System

A. Physicochemical:	6 ml of medium in #13 S.C. tubes leaves 34 percent air space.
	7 mg percent of 1-cysteine, HCL.

B. Components:

EK = Base salts +	%	NC – 5 = Base + supplements below	%
Na-glutamate	0.90	α-ketoglutaric acid	0.120
Na-pyruvate	0.30	Cytochrome C	0.012
Glucose	0.25	Hemin	0.003
Glycerol	2.00	1-cysteine, HCL	0.007
Ca-pantothenate	0.006	Serum	10.0/vol

pH = 7.0	pH of all = 7.0
Autoclave 121° – 15 min.	All filtered through
pH = 7.0	Millipore MF, 0.22 μm

Inoculate and immediately distribute 6 ml/tube.
All percentages = final concentrations in the medium.

Figure 2A demonstrates why quantitation of functional biomass (F Bm) is more instructive than microscopic counts or, indeed, plate counts of microbial species that possess a high plating efficiency. When the in vivo grown cells of *M. lepraemurium* are inoculated in the Kirchner semisynthetic base (EK), the F Bm of *M. lepraemurium* fell to 50 percent of the original at the third day

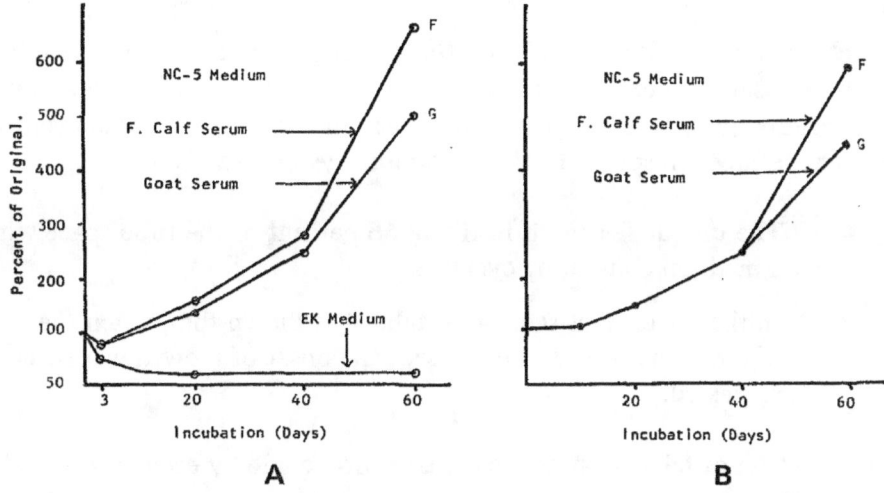

Figure 2. A—ATP per culture. B—Biomass per culture.

and declined progressively thereafter. This curve duplicates the slopes shown earlier with the mouse brain extract medium. Meanwhile, in the presence of the physicochemical conditions and catalysts provided by the supplements in the Nc-5 medium, the F Bm falls only to 75 percent of the original, that is, one half as far. Thus, within three days, it was known that the critical conditions and catalysts of the Nakamura system resided in the supplements, not in the base. By the 10th day in the complete medium (Nc-5 curve), the cultures had reestablished the energy levels utilized by *M. lepraemurium* for growth in mouse tissues, that is, the cells had successfully reconstructed the leaky in vivo type membranes into those which are competent for extracellular growth. On the NC curve between the 4th and the 10th day, *M. lepraemurium* already had spent 6 days in expanding its metabolic equipment and membrane activity and possibly had accomplished increases in cell volume.

Within 20 days the rate of expanding F Bm had improved slightly. After 20 days fetal calf serum is shown to be superior to goat serum. Thus, the goals of such experiments had been accomplished within three weeks.

After 10 days of lag (figure 2B), microscopic measurements of total biomass (cell numbers X their average length) increased in parallel with the ATP curves, but the cultures had to be incubated at least twice as long to guarantee that significant growth had occurred. Experience has shown that microscopic quantitations of the total Bm of *M. lepraemurium* cultures is exceptionally tedious. *M. lepraemurium* is one of the mycobacterial species that elongate during the early stages of growth. If the work required to enumerate bacterial cells by the pin head method is regarded as one unit, determination of the average length of the cells requires three units of effort. In short, valid microscopic measurements of the biomass of *M. lepraemurium* per culture requires not one, but four units of labor.

Physiochemically, *M. lepraemurium* hitherto has proven to be noncultivable in many thousands of experiments of dozens of investigators. Any system which permits the growth of such an organism in vitro will be nonconventional in some definable respect. In the Nakamura system (see table 1):

- The unique feature is in filling 66 percent of the tube space with a medium containing cysteine.

- In the lower right section of table 1 is shown that, when the ratio of air in the tube was increased, the onset of growth was severely repressed.

The left section of table 1 ranks the contribution made by each compound in the medium.

In order to facilitate investigations with host-grown microbes we outline two propositions:

1. Metabolic or biochemical indicators of physiologic states are more significant than microscopic or plate counts, because (a) such methods measure inhibitory conditions as readily as favorable ones, (b) they help to study the physiologic sag and thus the essentials can be learned without waiting for growth to occur, and (c) the lack of normal enzyme versatility makes host-dependent organisms particularly susceptible to inhibitions.

2. Given useful sources of both nitrogen and carbon and relatively constant conditions, ATP per culture measures F Bm, the parameter that should correlate with plate and microscopic counts.

Physiologic monitors are required to analyze the transition of in vivo-grown microbial cells from noncultivable to cultivable state. Even assuming the investigation of any microbe with high plating efficiency, plate or microscope counts cannot analyze the crucially important events which are fundamental to the adaptation of a host-dependent microbe to existance in vitro. Thus, the physiologic sag of the host-dependent microbe has been defined as analogous to, but far more severe than, the sag or lag which occurs when minimal inoculums of cultivable microbes are transferred to a new medium. The observations here demonstrated that measurement of F Bm produces information of the type that cannot be gained by plate or microscopic counts, also that significant results can be obtained during half the incubation time required to obtain the earliest information by means of biologic methods.

In conclusion, I wish to summarize two points.

1. It appears that ATP measurements have been refined to a point where the growth potential of microscopic samples of so-called

Table 1

The Merits of Conditions and Components
of the Nakamura System

Item	% of NC-5	Rank	Item	% of NC-5	Rank
NC-5	100				
α-Ketoglutaric	62	1	Glycerol	6	7
Hemin	59		Glucose	5	
Cytochrome C	41	2	Depth of medium	% of air space	% of NC-5
FCS	40				
1-cysteine, HCl	29	3	6 ml	34	100
Na-pyruvate	24	4	5 ml	45	88
Ca-pantothenate	18	5	3 ml	67	52
Na-glutamate	15	6	1 ml	89	32

Method: Starting with the complete NC-5, each item was deleted in turn.

% of NC-5 = the percentage gain when each item was restored to the deficient medium.

host-dependent microbes can be measured, even before an increase in microbial numbers has occurred.

2. This tool differs from other biochemical indicators of physiologic states in its sensitivity and in the fact that it can be exploited with unwashed host-dependent and host-grown organisms. We trust that the present studies have created new means of investigating the energetic integrity and the biosynthetic potential in *M. leprae* during experiments of the type demonstrated and also during the progression, regression, and therapy of the disease in humans.

APPLICATIONS OF ADENINE NUCLEOTIDE MEASUREMENTS IN OCEANOGRAPHY

O. Holm-Hansen, R. Hodson, and F. Azam
Institute of Marine Resources
University of California at San Diego
La Jolla, California

INTRODUCTION

All our oceanographic studies involving measurement of ATP, ADP, or AMP are based on the following two premises.

- All live microbial cells in our samples contain a relatively uniform amount of ATP per unit cell volume. As dead cells and detrital material do not contain any significant amount of ATP, the measurement of ATP permits estimation of total microbial biomass in terms of weight or cellular organic carbon.

- The cellular concentrations of adenine nucleotides are important in regulation of metabolic rates. The relative concentrations of these nucleotides thus can yield information on the physiological state of the cells.

Before discussion of our specific uses of data on the distribution of these nucleotides, we wish to briefly outline (1) the methodology involved in these measurements and (2) our data to support the premise that ATP concentrations in microbial cells can be extrapolated to biomass parameters.

METHODOLOGY

Sampling

In most of our work, the concentration of cells is so low that we must concentrate the cells prior to killing and extraction for ATP. Our sampling devices range from small sterile units (for example, Niskin disposable bags) to large 200-l Van Dorn type samplers which are scrubbed with alcohol just prior to use. The interval between collection of the sample and killing of cells is kept as short as possible to minimize any changes in cellular ATP levels due to changes in pressure, temperature, or light conditions.

Killing of Cells and Extraction of Nucleotides

Cells are concentrated by filtration of the sample through microfine glass fiber filters (Reeve-Angel, No. 984H) which are then quickly immersed in TRIS buffer (0.02 M, pH 7.7) at 373 K (100° C). It is important to kill the cells and to stop all enzymatic activity as quickly as possible. The turnover time of ATP is very rapid (from about 1 s to 1 min) in most cells, and a delay in killing of the cells or a slow killing procedure can result in significant changes in ATP levels. After boiling for 4 to 5 min, the samples are stored at 253 K (-20° C) until time of analysis.

Measurement of ATP

During the past nine years, we have developed the necessary instrumentation to give us the speed, sensitivity, and flexibility that are required in our type of work. In most of our studies we measure ATP in the sample by integration of the light flux for a 1-min time period (figure 1). One important advantage of this method is that it permits one to obtain complete mixing of the sample with enzyme preparation during the 15-s delay period. It also eliminates the artifacts often associated with increased light levels due to agitation of the enzyme preparation when it is mixed with the sample. The magnitude of this latter problem varies considerably with different enzyme preparations.

For some research problems, the use of peak-height analysis offers some advantages (for example, time required per assay in analyses for ADP and AMP when enzymes other than luciferase are reacting with ATP) over the light-integration methods. For such studies we use another machine (JRB ATP Photometer) which displays either the integrated light flux or the peak-height value directly on a digital readout. The sensitivity of these photometers with partially purified enzyme is about 10^{-16} moles ATP. For a more detailed description of the methodology, see Holm-Hansen (1973).

ATP CONCENTRATIONS IN MICROORGANISMS

For our purposes the most useful parameter to use for biomass estimation is total cellular organic carbon. During the past nine years, we have studied a great variety of marine and freshwater microbial cells to obtain data on the ratio of cell carbon to ATP, and how this ratio can be altered by environmental stresses. The carbon per ATP ratio we use in our work is 250, which is based on the following types of data.

Figure 2 shows the relationship between cellular organic carbon and ATP in 30 species of unicellular algae representing 7 phyla. Details on these cultures have been described by Holm-Hansen (1970a).

Hamilton and Holm-Hansen (1967) used both batch and chemostat cultures of marine bacteria and measured ATP, total cell count, viable cell count, and

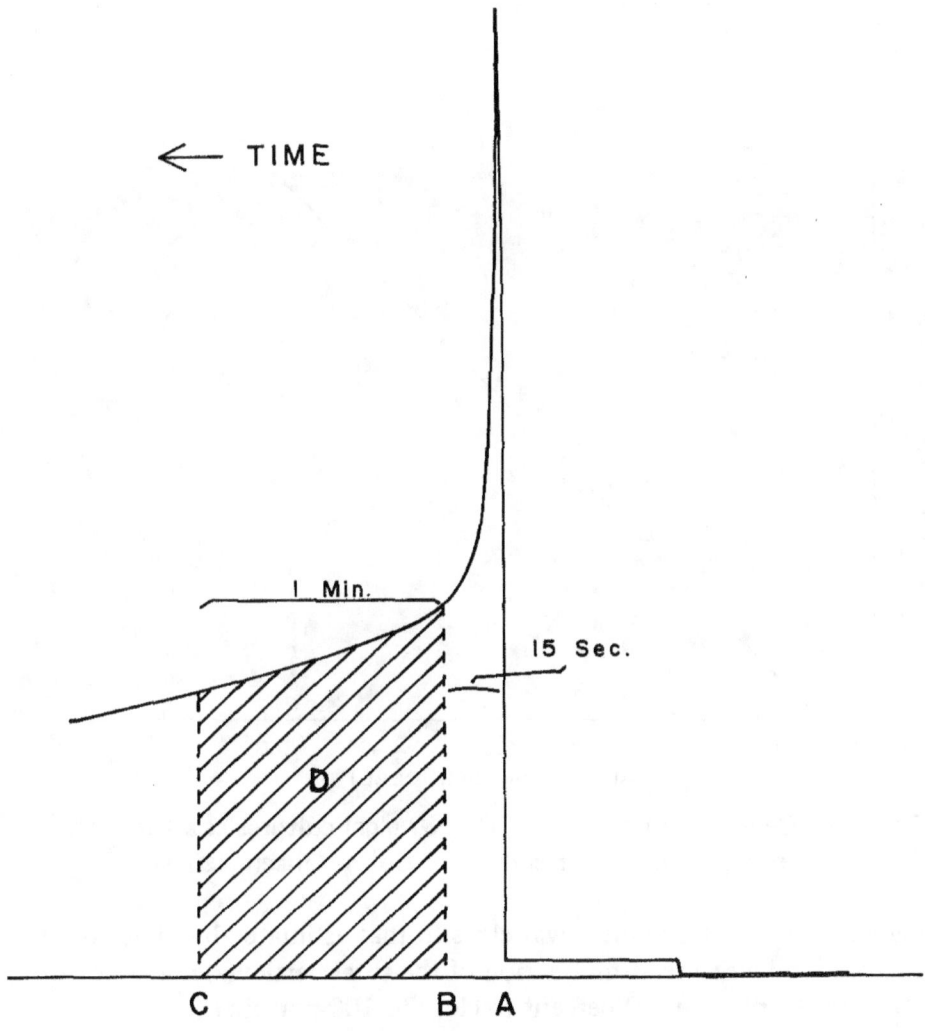

Figure 1. Light emission curve when a sample containing ATP is injected into the luciferin-luciferase enzyme preparation: A—time of injection of sample; B—end of 15-s mixing period; C—end of 1.0-min light-integration period; D—area electronically integrated (from Holm-Hansen, 1973).

cellular organic carbon. These data showed that ATP was a measure of the concentration of viable cells in the suspension and that the ratio of carbon to ATP averaged close to the above value of 250.

We have also made numerous measurements of ATP on natural phytoplankton populations, either with or without nutrient enrichment (Strickland et al., 1969; Eppley et al., 1971, Holm-Hansen, 1969). In these studies we have also measured total particulate organic carbon in addition to estimating phytoplankton-carbon by calculations based on cell counts and cell volumes as measured with an inverted microscope. In all this work, the carbon as

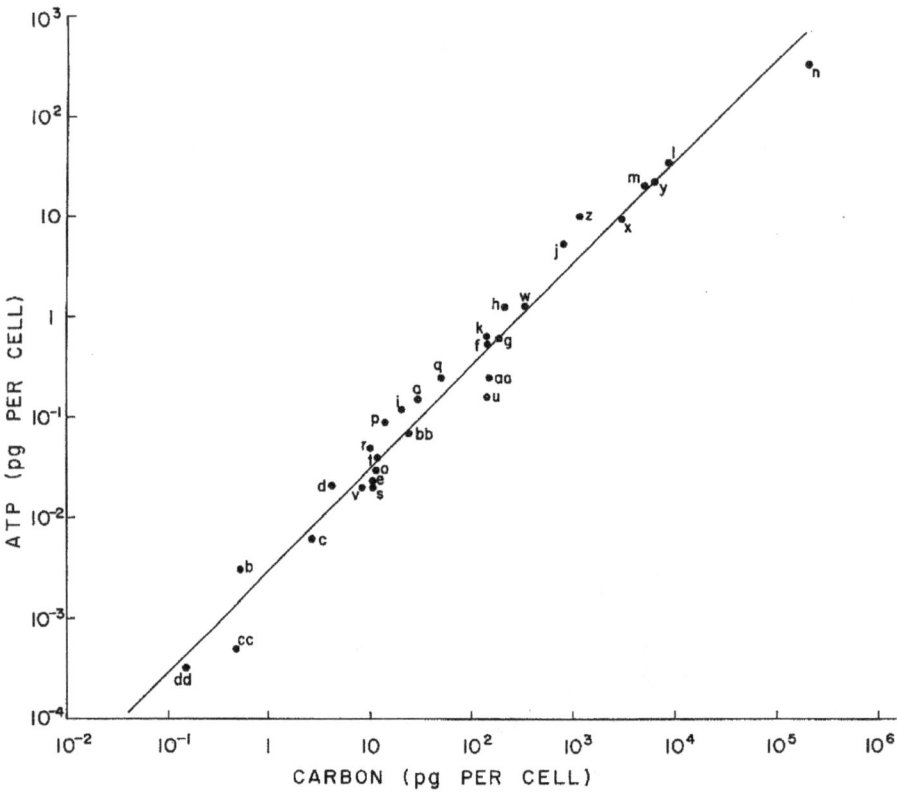

Figure 2. Cellular contents of ATP in 30 algal cultures as a function of
the organic carbon per cell (from Holm-Hansen, 1970a).

estimated by ATP measurement was close to that estimated by direct floristic analyses, and was a realistic fraction of the total particulate organic carbon (ranging from about 20 percent to close to 100 percent).

As cells are taken from many depths in the water column, they will be exposed to varying light conditions during subsampling and filtration. It was therefore of importance to our field work to ascertain what changes in light conditions will do to cellular ATP levels. Holm-Hansen (1973) has shown that ATP levels in phytoplankton are maintained at fairly uniform concentrations during light/dark periods. There are transient changes in ATP levels when light conditions are suddenly changed, but the levels are reestablished within a period of 1 to 3 min.

Both microzooplankton (flagellated or ciliated protozoans) and macrozooplankton (Crustacea) also have ATP contents very close to 0.4 percent of the total cellular organic carbon. Some data for *Calanus* species are included in the paper by Holm-Hansen (1973).

During the past year, Thiel and Holm-Hansen have been studying the ATP content of various benthic animals, from both littoral and deep (1200 m) sediments. Our samples have included representatives of the Annelida,

Platyhelminthes, Aschelminthes, and Arthropoda (Crustacea). There is considerably more spread in the C/ATP ratio in these animal phyla, as the ratio in some polychaetes has been close to 100, while in some amphipods it is close to 1000. These latter individuals, however, are large (2 to 3 mg dry weight) and the exoskeleton probably accounts for close to 75 percent of the total cellular organic carbon. Most of our data on these diverse animals show carbon per ATP (C/ATP) ratios between 200 to 300. We therefore believe that the relationship of cell carbon to ATP as used for all our phytoplankton work will also be applicable to bottom-dwelling metazoans.

APPLICATIONS OF NUCLEOTIDE ANALYSES

Field Applications

One of the primary applications for which we have used ATP determinations involves questions concerning the distribution of microbial cells throughout the entire water column. The primary input of reduced carbon into the marine food chain is via phytoplankton photosynthesis, which is limited to the upper 150 m or less of the water column. We have considerable data on the distribution of larger zooplankton and fish throughout the water as well as on the benthic organisms, but we do not know the source of food necessary to support these populations. Two possibilities which are often discussed in this context are: (a) that there is an active transfer of particulate carbon from the euphotic zone throughout the entire water column by migrating zooplankton populations (Vinogradov, 1962a), and (b) that there is a significant population of heterotrophically growing cells in deep water which are grazed by filter-feeding organisms (Holm-Hansen, 1970b).

Until the introduction of ATP analyses to oceanographic work (Holm-Hansen and Booth, 1966), microbial biomass estimates were obtained either by agar-plating techniques or by direct microscopic examinations. Both of these methods have serious limitations, especially when working with deep samples (Jannasch and Jones, 1959). Figure 3 shows the microbial biomass estimated by ATP analyses to a depth of 4300 m at a station in the north Pacific. Microbial biomass decreases from 23 μg carbon in the euphotic zone (where the total particulate organic carbon was about 30 μg/l) to values of about 0.1 μg C/l (total particulate organic carbon (POC) being about 3 to 5 μg C/l). Also shown on figure 3 are the data which Fournier (1971) obtained for total biomass of yellow-green cells from the same water samples as used for our ATP analyses. The significance of these pigmented cells in deep water is not known, although Fournier (1973) has described data which he interprets as showing that these cells may be important in deep-sea food chains. It should be noted that the biomass of microbial cells as indicated by ATP analyses is in the order of 10 times the biomass of zooplankton from similar depths as described by Vinogradov (1962b).

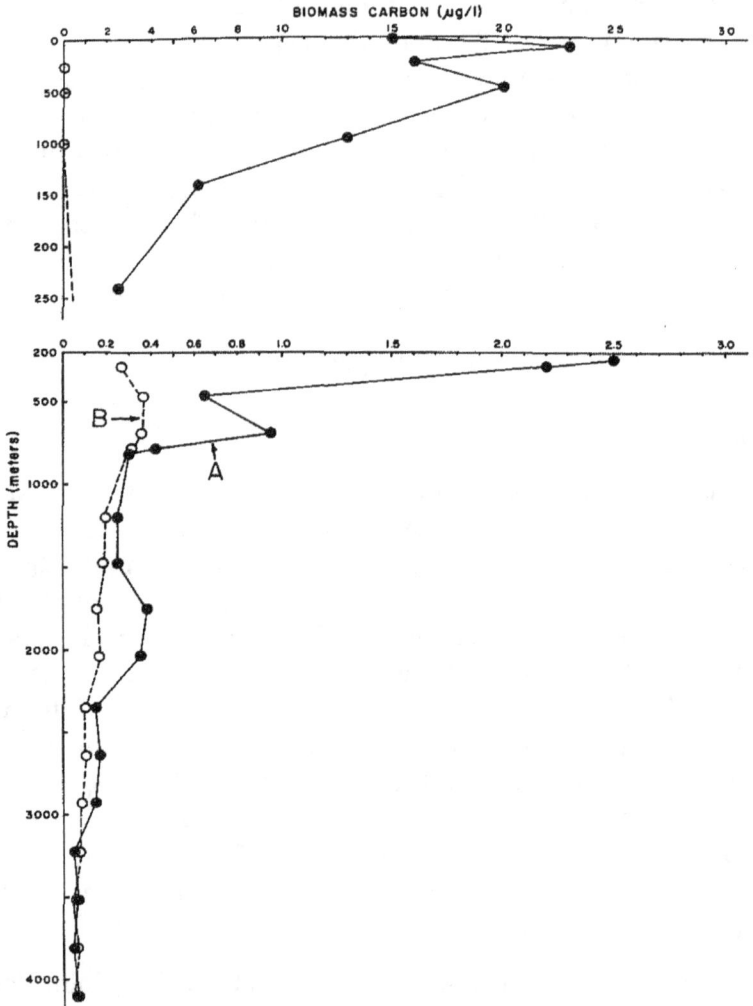

Figure 3. The distribution of microbial biomass-carbon as determined from ATP measurements (line A) and the calculated biomass-carbon contained in olive-green cells (line B) from Station Gollum north of Oahu (station position: 22°10'N, 158°00'W). The data for the olive-green cells have been taken from the paper by Fournier (1971) (from Holm-Hansen and Paerl, 1972).

The same types of problems as discussed above are also of much interest in regard to organisms living in or on bottom sediments. We are now applying ATP methodology to estimate microbial biomass in marine sediment profiles. This work is part of a larger program directed by Dr. Hessler (Scripps), which is concerned with the distribution and activity of both microbial and metazoan species in deep sea sediments.

In addition to merely assessing the standing crop of microbial cells, we have also used ATP determinations in various studies concerned with the metabolic activity of microbial cells. Two such examples are described below.

80

Figure 4 shows the distribution of ATP in the extremely oligotrophic waters of the north Pacific gyre region. The general characteristics of the water

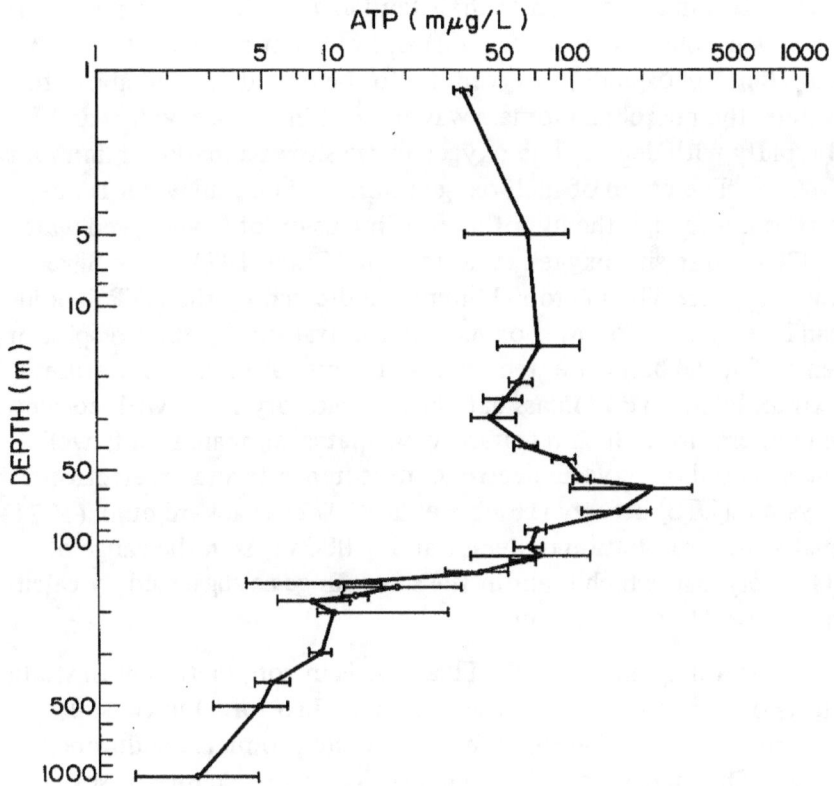

Figure 4. Vertical distribution of ATP in the north Pacific gyre. Station position: 28°00'N, 155°00'W.

column at this location have been described by Venrick et al., (1973) and Eppley et al. (1973). The phytoplankton biomass is low in the upper 50 m and generally reaches a maximum at about 100 to 140 m. Nutrients are very low in the upper 100 m, with nitrate and phosphate generally being close to or below the detection limits of our methods (about 0.02 μM) and silicate being about 1.0 μM. The 1 percent light levels (the conventional compensation depth for algal photosynthesis) is usually between 80 and 140 m. The productivity profile, however, usually shows a maximum in the upper 50 m, with decreasing rates below that depth. Phytoplankton productivity in these waters thus will be nutrient-limited in the upper portion of the euphotic zone and light-limited in the lower portions. The rate of input of inorganic nutrients into the euphotic zone will therefore be one of the prime factors influencing the rate of primary production. It can be seen from figure 4 that there is a substantial microbial population between 100 and 1000 m, and it can be assumed that these cells are dependent mostly upon heterotrophic processes for their carbon and energy sources. These microbial populations are likely to be of considerable importance in regard to regeneration of nutrients which are essential for continued phytoplankton productivity.

Figure 5 shows a profile for ATP and dissolved oxygen concentrations from the surface to 7200 m at a station in the Aleutian Trench. The surface water at this station was cold (280.9 K (7.8° C)) and nutrient-rich (about 15 μM silicate and 1.0 μM phosphate). In contrast to the data discussed above for tropical waters, the microbial biomass was greatest in surface waters and decreased rapidly with depth. The oxygen curve shows a marked minimum at about 900 m. The origin of such oxygen minima in ocean water is not well understood, although the hypotheses of intrusion of low-oxygen water (Menzel, 1970) and in situ oxygen consumption (Craig, 1971) both have considerable support. The microbial biomass indicated by the ATP data in figure 5 can be expected to lower oxygen concentrations by their respiratory requirements. On the basis of extensive respiratory data in the literature we have extrapolated ATP-biomass data into respiratory rates, with corrections for in situ temperatures. Such respiratory estimates compared quite well with estimates based on oxygen electrode measurements and on electron transport system (ETS) activity (Hobbie et al., 1972). Packard et al. (1971) have estimated that respiration in deep water (5000 m) is in the range of 1 to 5 μl O_2/liter/year, which is about the same range as suggested by calculations based on ATP measurements.

A particularly useful application of ATP assays is in conjunction with studies on the effects of outfalls (sewer, nuclear plant, and so forth) in coastal waters or on the effects on benthic life when waste products are dumped in deep water. The increase of microbial biomass in the vicinity of sewer outfalls has been documented by Eppley et al. (1972). Similar studies on the biomass of microbial cells in sediments would be of considerable interest, as many pollutants will have limited solubility in water and will concentrate in the sediments.

Some pollutants apparently can act either as killing agents or as growth-inhibiting agents. Fitzgerald and Faust (1963), for instance, have shown that Cu displays either algicidal or algistatic properties on various species of algae. ATP measurements can be used to differentiate between dead cells and such inhibitor-arrested cells. We are currently using this technique in our controlled ecosystem pollution experiment (CEPEX) program, in which we are examining the effects of various pollutants on natural phytoplankton populations.

Laboratory Studies

One of our main projects is the study of the factors which control or limit phytoplankton growth in the sea. Reference to figure 6 will indicate the nature of one problem in interpretation of our ^{14}C-uptake studies. In primary production experiments using ^{14}C-bicarbonate, the incubation period normally used ranges from a few hours to one day. When cells are very stressed by nutrient deficiency, however, there can be a 1- to 2-day lag after introduction of the limiting nutrient before reduction of CO_2 occurs at any

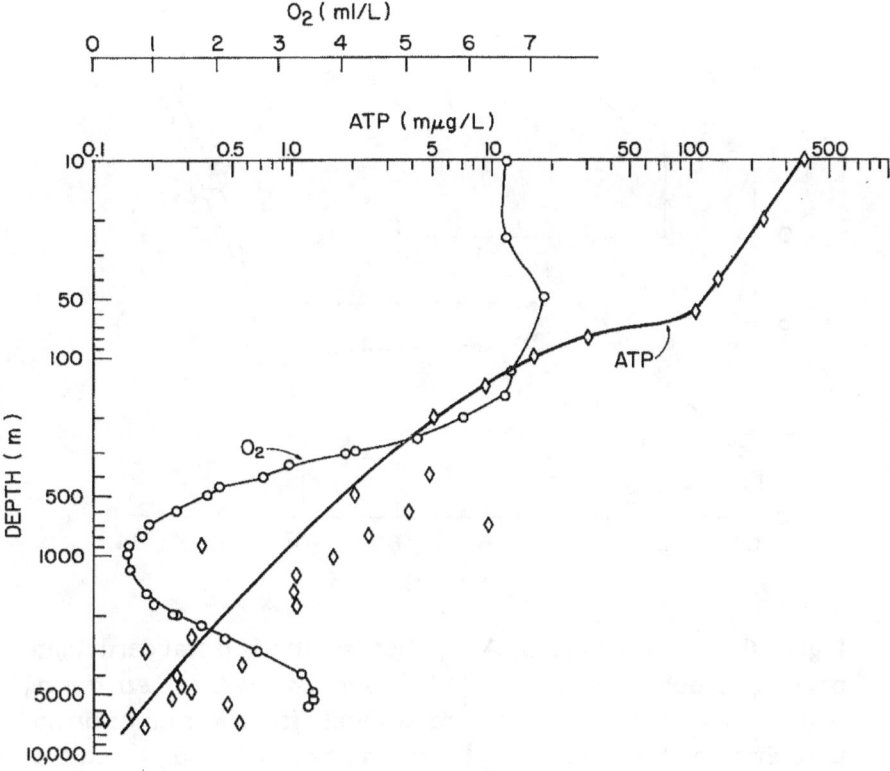

Figure 5.   Vertical distribution of ATP and dissolved O₂ in the Aleutian Trench. Station position: 50°53'N, 177°31'W.

appreciable rate. There is, however, a rapid response of cellular ATP levels upon addition of the nutrient (see inset of figure 6). The chemical composition of cells during starvation and after recovery can be very different. After addition of phosphorus to P-deficient cells, for instance, the C/P ratios may change from 250 to 5 and the C/ATP ratio from 2000 to 200. Such studies on the biochemical events occurring during the period when cells are adjusting to nutrient addition are useful in conjunction with our biomass estimations based on ATP concentrations and with assessment of the nutritional status of phytoplankton populations in nature.

Most blue-green algae are obligate photoautotrophs and cannot grow heterotrophically in the dark. Smith et al. (1967) have suggested that the reason for this is that these algae lack the complete complement of tricarboxylic acid (TCA) cycle enzymes and hence are incapable of oxidative phosphorylation. They would thus depend solely upon photophosphorylation for their ATP production. Our data on ATP levels in the blue-green alga *Nostoc muscorum*, however, demonstrate that this hypothesis of Smith et al. is not correct. As in all other algae investigated, ATP levels decrease rapidly when transferred from light to dark, but within a few minutes oxidative phosphorylation restores the ATP levels to the original light level.

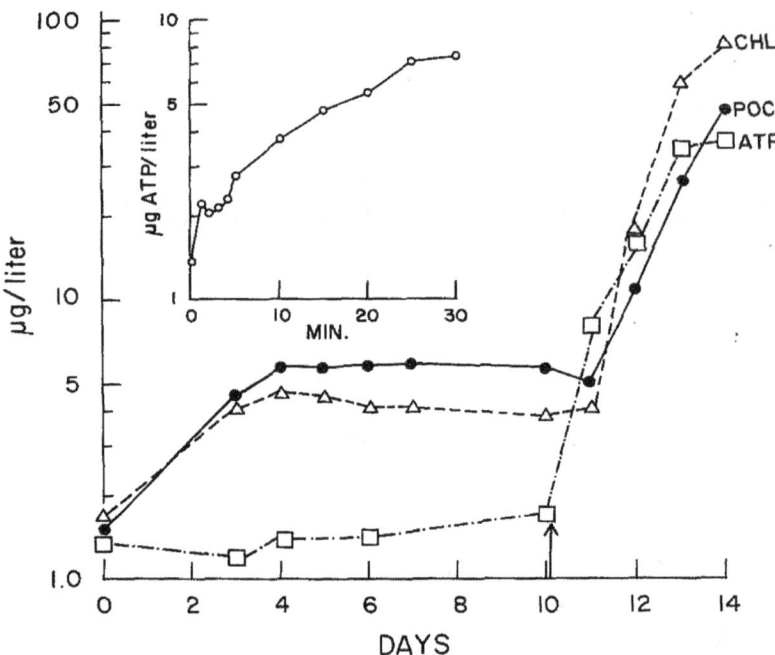

Figure 6. The response of ATP, photosynthesis (total particulate organic carbon), and chlorophyll in *Monochrysis lutheri* during phosphorus deficiency (days 1 to 10) and after addition of orthophosphate at time indicated by the arrow. ATP, $\mu g/l$; chlorophyll, $\mu g \times 10^{-1}/l$; POC, $\mu g \times 10^{-3}/l$. Inset shows ATP response from 1 to 30 minutes in a separate experiment when phosphate was added to deficient cells at time 0.

Silicon is an essential mineral element for most diatoms. When a culture of *Navicula pelliculosa* is silicon-deficient, mitosis occurs without concomitant cytokinesis (Coombs et al., 1967a). A population of binucleated cells thus results, which remains in this condition until addition of silicon. When silicon is introduced, there are rapid changes in respiration, photosynthesis, and assimilation of silicon (Coombs et al., 1967b). Transient changes in ATP levels have been followed throughout this sequence of events in order to see if the uptake of silicon is an active uptake mechanism which requires ATP. The results of these investigations have been described by Coombs et al. (1967a).

As we have seen from the other papers in this symposium, the energy charge of cells as described by Atkinson (1969) does seem to be correlated with viability and growth rate of cells. We have therefore been investigating the relative concentrations of ATP, ADP, and AMP in laboratory cultures of stressed cells discussed above. Under extreme nitrogen or phosphorus nutrient deficiency, the ratio of C/ATP increases significantly, which would result in an error in biomass estimations which assume a constant ratio of C/ATP.

We are hopeful that in such situations, measurement of all three nucleotides will be more useful than measurement of ATP alone.

In an earlier section we discussed estimation of microbial biomass in deep water and in sediments. In addition to standing crop estimates, we also need information on the metabolic activity of these cells, which requires use of other techniques such as heterotrophic uptake studies with radio-labeled substrates. We are now also examining the energy charge of these cells living in deep water in the hope that these data will be informative in regard to the energetic state of the cells.

ACKNOWLEDGMENTS

This work was supported in part by AEC Contract AT(11-1) GEN 10, P.A. 20 and in part by NSF Grants GV-41705 and GX-39139.

We thank Dr. P. M. Williams for permission to use data collected on Expedition Seven-Tow, Leg 7.

REFERENCES

Atkinson, D. E. "Regulation of Enzyme Function," *Ann. Rev. Microbiol.,* **23**, 1969, p. 47-68.

Coombs, J., P. J. Halicki, O. Holm-Hansen, and B. E. Volcani, "Studies on the Biochemistry and Fine Structure of Silica Shell Formation in Diatoms. II. Changes in Concentrations of Nucleoside Triphosphates in Silicon-starvation Synchrony of *Navicula pelliculosa* (Breb.) Hilse," *Exper. Cell Res.,* **47**, 1967a, p. 315-328.

Coombs, J., C. Spanis, and B. E. Volcani, "Studies on the Biochemistry and Fine Structure of Silica Shell Formation in Diatoms. Photosynthesis and Respiration in Silicon-starvation Synchrony of *Navicula pelliculosa,*" *Plant Physiol.,* **42**(11), 1967b, p. 1607-1611.

Craig, H., "The Deep Metabolism: Oxygen Consumption in Abyssal Ocean Water," *J. Geophys. Res.,* **76**, 1971, p. 5078-5086.

Eppley, R. W., A. F. Carlucci, O. Holm-Hansen, D. Kiefer, J. J. McCarthy, E. Venrick, and P. M. Williams, "Phytoplankton Growth and Composition in Shipboard Cultures Supplied with Nitrate, Ammonium, of Urea as the Nitrogen Source," *Limnol. Oceanog.,* **16**(5), 1971, p. 741-751.

Eppley, R. W., A. F. Carlucci, O. Holm-Hansen, D. Kiefer, J. J. McCarthy, and P. M. Williams, "Evidence for Eutrophication in the Sea Near Southern California Coastal Sewage Outfalls—July 1970," Calif. Mar. Res. Comm., CalCOFI Rept., **16**, 1972, p. 74-83.

Eppley, R. W., E. H. Renger, E. L. Venrick, and M. M. Mullin, "A Study of Plankton Dynamics and Nutrient Cycling in the Central Gyre of the North Pacific Ocean," *Limnol. Oceanog.*, **18**, 1973, p. 534-551.

Fitzgerald, G. P. and S. L. Faust, "Factors Affecting the Algicidal and Algistatic Properties of Copper," *Appl. Microbiol*, **11**, 1963, p. 345-351.

Fournier, R. O., "Studies on Pigmented Microorganisms from Aphotic Marine Environments. II. North Atlantic Distribution," *Limnol. Oceanog.*, **16**, 1971, p. 952-961.

Fournier, R. O., "Studies on Pigmented Microorganisms from Aphotic Marine Environments. III. Evidence of Apparent Utilization by Benthic and Pelagic Tunicata," *Limnol. Oceanog.*, **18**, 1973, p. 38-43.

Hamilton, R. D., and O. Holm-Hansen, "Adenosine Triphosphate Content of Marine Bacteria," *Limnol. Oceanog.*, **12**(2), 1967, p. 319-324.

Hobbie, J. E., O. Holm-Hansen, T. T. Packard, L. R. Pomeroy, R. W. Sheldon, J. P. Thomas, and W. J. Wiebe, "A Study of the Distribution and Activity of Microorganisms in Ocean Water," *Limnol. Oceanog.*, **17**(4): 1972, p. 544-555.

Holm-Hansen, O., "Determination of Microbial Biomass in Ocean Profiles," *Limnol. Oceanog.*, **14**(5), 1969, p. 740-747.

Holm-Hansen, O., "Determination of Total Microbial Biomass by Measurement of Adenosine Triphosphate," *Estuarine Microbial Ecology*, L. H. Stevenson and R. R. Colwell, eds., University of South Carolina Press, 1973.

Holm-Hansen, O. and H. W. Paerl, "The Applicability of ATP Determination for Estimation of Microbial Biomass and Metabolic Activity," *Mem. Ist. Ital. Idrobiol.*, **29**, Suppl., 1972, p. 149-168.

Holm-Hansen, O., "ATP Levels in Algal Cells as Influenced by Environmental Conditions," *Plant and Cell Physiol.*, **11**, 1970a, p. 689-700.

Holm-Hansen, O., "Microbial Distribution in Ocean Water Relative to Nutrients and Food Sources," *Biological Sound Scattering in the Ocean*, G. B. Farquhar, ed., Maury Center for Ocean Science, Washington, D. C., 1970b.

Holm-Hansen, O. and C. R. Booth, "The Measurement of Adenosine Triphosphate in the Ocean and Its Ecological Significance," *Limnol. Oceanog.*, **11**, 1966, p. 510-519.

Jannasch, H. W. and G. E. Jones, "Bacterial Populations in Sea Water as Determined by Different Methods of Enumeration," *Limnol. Oceanog.*, **4**, 1959, p. 128-139.

Menzel, D. W., "The Role of in situ Decomposition of Organic Matter on the Concentration of Nonconservative Properties in the Sea," *Deep-Sea Res.,* **17**, 1970, p. 751-764.

Packard, T. T., M. L. Healy, and F. A. Richards, "Vertical Distribution of the Respiratory Electron Transport System in Marine Plankton," *Limnol. Oceanog.,* **16**, 1971, p. 60-70.

Smith, A. F., J. London, and R. Y. Stanier, "Biochemical Basis of Obligate Autotrophy in Blue-green Algae and Thiobacilli," *J. Bacter.,* **94**, 1967, p. 972-983.

Strickland, J. D. H., O. Holm-Hansen, R. W. Eppley, and R. J. Linn, "The Use of a Deep Tank in Plankton Ecology. I. Studies of the Growth and Composition of Phytoplankton Crops at Low Nutrient Levels," *Limnol. Oceanog.,* **14**(1), 1969, p. 23-34.

Venrick, E. L., J. A. McGowan, and A. W. Mantyla, "Deep Maxima of Photosynthetic Chlorophyll in the Pacific Ocean," *Fish. Bull.,* **71**(1), 1973, p. 41-52.

Vinogradov, M. E., "Quantitative Distribution of Deep-sea Plankton in the Western Pacific and Its Relation to Deep-sea Circulation," *Deep-Sea Res.,* **8**, 1962a, p. 251-258.

Vinogradov, M. E., "Feeding of the Deep-sea Zooplankton," *Rapp. Cons. Perm. Int. Explor. Mer,* **153**, 1962b, p. 114-120.

SPECIFICITY OF AEQUORIN LUMINESCENCE TO CALCIUM

Osamu Shimomura and Frank H. Johnson
Biology Department
Princeton University
Princeton, New Jersey

SUMMARY

The presence of Pb^{++}, Co^{++}, Cu^{++}, and Cd^{++}, each of which possesses a certain luminescence-triggering activity of aequorin, potentially interferes with the specificity of the aequorin luminescence response to Ca^{++}. Interference by the above cations can be eliminated, without influencing the sensitivity of the luminescence of aequorin to Ca^{++}, by adding 1 mM of sodium diethyldithiocarbamate.

INTRODUCTION

The isolation of a substance involved as an active principle in the luminescence of *Aequorea aequorea*, a hydromedusan jellyfish indigenous to coastal waters of the Pacific Northwest, was reported some years ago (Shimomura et al., 1962). The purified substance proved to be a conjugated protein for which the name aequorin was introduced. Later studies on other bioluminescent organisms, *Chaetopterus* and *Meganyctiphanes* (Shimomura and Johnson, 1966; 1968), indicated that aequorin was the first example of a type of luminescence system that was given the collective name "photoprotein," characterized by the properties that (1) under optimum conditions the total amount of light emitted was proportional to a specific protein, and (2) the light-emitting process did not directly involve any enzyme, at least in the usual sense of the term.

In aqueous solution, aequorin was found to have the unique property of being triggered by traces of Ca^{++}, and to a lesser extent by Sr^{++}, to emit light, either with or without molecular oxygen. The sensitivity of the luminescence response to Ca^{++} is estimated to extend to as little as a small volume of 10^{-7} M solution (Shimomura et al., 1963a; Shimomura and Johnson, 1973). The light-emitting reaction seems not susceptible to inhibition by any compounds commonly occurring in biological systems. Moreover, no evidence has been found that aequorin has any toxic properties. The speed of response to Ca^{++} is fast; maximum intensity is reached in a few milliseconds (Hastings et al., 1969; Loschen and Chance, 1971).

On the basis of sensitivity together with the favorable properties just mentioned and a seemingly satisfactory specificity, the use of aequorin luminescence as a means of the microdetermination of Ca^{++} in biological systems was suggested (Shimomura et al., 1963b). The method was soon applied to advantage in the detection of changes in Ca^{++} concentration during the contraction of single muscle fibers (Ridgway and Ashley, 1967; Ashley and Ridgway, 1968), and, in a number of other physiological processes including metabolic activity of mitochondria (Azzi and Chance, 1969), activity of single neurons as well as action of inhibitors thereon (Baker et al., 1971), and rhythmic discharge of single ganglionic cells (Chang et al., 1974). In studies of Ca^{++} inside single cells, one would be most likely dealing with 10^{-17} to 10^{-15} mole ions of ionic calcium, assuming a cell size of 100 μm in diameter.

In regard to specificity, extended research on the influence of various cations (Izutsu et al., 1972; Shimomura and Johnson, 1973) have shown that, under certain conditions of concentration and pH, cations other than Ca^{++} and Sr^{++} can trigger the luminescence of aequorin. Such cations include those of the rare earth elements and also Pb^{++}, Cd^{++}, Co^{++}, and Cu^{++}. In the present study, we tried to improve further the specificity of aequorin reaction of Ca^{++} by masking these cations other than Ca^{++} through the use of various chelating agents.

MATERIALS AND METHODS

Purified aequorin of approximately 95 percent purity (Shimomura and Johnson, 1969; Johnson and Shimomura, 1972) was desalted by a small column of Sephadex G-25 equilibrated with 0.1 mM ethylene diaminetetra acetate (EDTA), pH 7.5. The solution was kept frozen until ready for use. CH_3COONa, $Pb(CH_3COO)_2$, CuO, CoO, $Cd(HCOO)_2$, $LaCl_3$, $SrCl_2$ (all ultrapure grade) and Y_2O_3 (99.999 percent) were obtained from Alpha Inorganics, and all other chemicals were reagent grade, with care taken to choose those containing a minimum amount of Ca. Deionized distilled water having a resistance of more than 10 MΩ was used, and all solutions were prepared and kept in polypropylene containers from which they were dispensed with plastic pipets and were not allowed contact with glass throughout the experiments, except for dissolving the metal oxides with HCl in Vycor test tubes as previously described (Shimomura and Johnson, 1973).

The luminescence reaction was initiated by rapid addition of 4 ml of buffer solution containing the metal salt to 5 μl of aequorin solution containing 10 μg of aequorin placed in a polycarbonate test tube. Light (λ max 470 nm) was measured by a photomultiplier-amplifier-recorder assembly with a pen response time of approximately 20 ms for the full scale.

RESULTS

Among various chelating agents tested, sodium diethyldithiocarbamate (DDC) was found to be far more effective than 8-hydroxyquinoline, 8-hydroxyquinoline-5-sulfonic acid, salycylic acid, oxalic acid, or thiourea, in regard to the masking of Pb^{++}, Cd^{++}, Co^{++}, and Cu^{++} to improve specificity of the aequorin reaction to Ca^{++}. The data for DDC, summarized in table 1, indicate that 1 mM of DDC completely suppressed the activities of 0.1 mM each of Pb^{++}, CD^{++}, Co^{++}, and Cu^{++}, but had little effect on the activities of Ca^{++}, Sr^{++}, and the representatives of the rare earth metals, La^{+++} and Y^{+++}. Trials with a higher concentration of DDC (5 mM) resulted in a considerable decrease in the total light emitted with Ca^{++}, indicating the partial destruction of aequorin; thus, approximately 1 mM is considered to be the optimum concentration for DDC.

In the absence of DDC, the effects of heavy metal ions are often enhanced in the presence of thiol (SH) compounds. For example, Pb^{++} and Cd^{++}, each at a concentration of 10^{-4} M in a buffer solution (pH 8) containing 1 mM of cysteine or 1 mM of 2-mercaptoethanol, both induced a luminescence of 1.5×10^{12} photons/s. Such luminescence, however, could be completely suppressed by addition of 1 mM of DDC, the same as without SH compounds as shown in table 1.

The activities of 0.1 mM of La^{+++} and Y^{+++} were suppressed by 8-hydroxyquinoline-5-sulfonic acid (1 mM) to a level lower than the activity of Sr^{++}, although this reagent was found to be relatively ineffective in masking Pb^{++} and Cd^{++}.

In regard to the microdetermination of Ca^{++} by the luminescence of aequorin, in general this method evidently has some unique advantages from the points of view of speed and of harmlessness to biological materials. Furthermore, it is satisfactorily specific to Ca^{++} at a pH of 8 in the absence of rare earth elements and of high concentrations of Sr^{++}. Although the activity of Pb^{++}, Co^{++}, Cu^{++}, and Cd^{++} can be completely eliminated by adding DDC, the occurrence of these ions in biological systems is normally unlikely; if DDC is used, however, control tests for possible harmful effects are obviously required.

ACKNOWLEDGMENT

We thank the National Science Foundation and the Office of Naval Research for partial support of this research. Reproduction in whole or in part for any purpose of the United States Government is permitted.

Table 1

Influence of Sodium Diethyldithiocarbamate (DDC) on the Maximum Intensities* of Aequorin† Luminescence Induced by Various Metal Ions‡

Metal ion added (0.1 mM)	In 0.01 M Na-acetate pH 6.0		In 0.01 M glycylglycine-NaOH pH 8.0	
	Without DDC	With 1.0 mM DDC	Without DDC	With 1.0 mM DDC
None	<0.02	<0.02	<0.02	<0.02
Ca^{++}	36.	32.	40.	26.
Sr^{++}	4.	4.	9.	7.
Pb^{++}	15.	<0.02	1.§	<0.02
Co^{++}	0.05	<0.02	1.§	<0.02
Cu^{++}	0.05	<0.02	0.5	<0.02
Cd^{++}	18.	<0.02	0.05	<0.02
Y^{+++}	34.	28.	14.	12.
La^{+++}	24.	20.	20.	20.

* Expressed in 10^{12} photons/s. At 298 K (25°C).

† 10 μg aequorin (potentiality = 42 × 10^{12} photons) was used in each test.

‡ In the present tests, those cations which have been found inactive in eliciting a luminescence reaction of aequorin, namely, K^+, Be^{++}, Mg^{++}, Ba^{++}, Mn^{++}, Ni^{++}, Fe^{++}, Fe^{+++}, and Zn^{++} (Shimomura and Johnson, 1973) were omitted.

§ A momentary flash took place at the start of the reaction (1.5 × 10^{12} photons in 0.1 to 0.2 s).

REFERENCES

Ashley, C. C. and E. B. Ridgway, "Simultaneous Recording of Membrane Potential, Calcium Transient and Tension in Single Muscle Fibres," *Nature*, **219**, 1968, p. 1168-1169.

Azzi, A. and B. Chance, "The Energized State of Mitochondria: Lifetime and ATP Equivalence," *Biochim. Biophys. Acta*, **189**, 1969, p. 141-151.

Baker, P. F., A. L. Hodgkin, and E. B. Ridgway, "The Early Phase of Calcium Entry in Giant Axons of *Loligo*," *J. Physiol.*, **214**, 1971, p. 33-34.

Chang, J. J., A. Gelperin, and F. H. Johnson, "Intracellularly Injected Aequorin Detects Transmembrane Calcium Flux During Action Potentials in an Identified Neuron from the Terrestrial Slug, *Limax maximus*," *Brain Res.*, 77, 1974, p. 431-442.

Hastings, J. W., G. Mitchell, P. H. Mattingly, J. R. Blinks, and M. van Leeuwen, "Response of Aequorin Bioluminescence to Rapid Changes in Calcium Concentration," *Nature*, **222**, 1969, p. 1047-1050.

Izutsu, K. T., S. P. Felton, L. A. Siegel, W. T. Yoda, and A. C. N. Chen, "Aequorin: Its Ionic Specificity," *Biochem. Biophys. Res. Commun.*, **49**, 1972, p. 1034-1039.

Johnson, F. H. and O. Shimomura, "Preparation and Use of Aequorin for Rapid Microdetermination of Ca^{++} in Biological Systems," *Nature New Biol.*, **237**, 1972, p. 287-288.

Loschen, G. and B. Chance, "Rapid Kinetic Studies of the Light-Emitting Protein Aequorin," *Nature New Biol.*, **233**, 1971, p. 273-274.

Ridgway, E. B. and C. C. Ashley, "Calcium Transients in Single Muscle Fibers," *Biochem. Biophys. Res. Commun.*, **29**, 1967, p. 229-234.

Shimomura, O. and F. H. Johnson, "Partial Purification and Properties of the *Chaetopterus* Luminescence System," *Bioluminescence in Progress*, F. H. Johnson and Y. Haneda, eds., Princeton University Press, 1966, p. 495-521.

Shimomura, O. and F. H. Johnson, "*Chaetopterus* Photoprotein: Crystallization and Cofactor Requirements for Bioluminescence," *Science*, **159**, 1968, p. 1239-1240.

Shimomura, O. and F. H. Johnson, "Properties of the Bioluminescent Protein Aequorin," *Biochemistry*, **8**, 1969, p. 3991-3997.

Shimomura, O. and F. H. Johnson, "Further Data on the Specificity of Aequorin Luminescence to Calcium," *Biochem Biophys. Res. Commun.*, **53**, 1973, p. 490-494.

Shimomura, O., F. H. Johnson, and Y. Saiga, "Extraction, Purification and Properties of Aequorin, a Bioluminescent Protein from the Luminous Hydromedusan *Aequorea*," *J. Cell. Comp. Physiol.*, **59**, 1962, p. 223-239.

Shimomura, O., F. H. Johnson, and Y. Saiga, "Further Data on the Bioluminescent Protein, Aequorin," *J. Cell. Comp. Physiol.*, **62**, 1963a, p. 1-8.

Shimomura, O., F. H. Johnson, and Y. Saiga, "Micro-determination of Calcium by Aequorin Luminescence," *Science*, **140**, 1963b, p. 1139-1140.

APPLICATIONS OF CHEMILUMINESCENCE TO BACTERIAL ANALYSIS

Norma D. Searle
American Cyanamid Company
Bound Brook, New Jersey

A rapid chemical method for determining bacteria populations would have obvious advantages over the 2-day microbiological plate count method in many areas of biological testing. Therefore a study was undertaken to investigate the use of the luminol chemiluminescence method for rapid detection of bacteria in a variety of applications. Previous studies of the method (Oleniacz et al., 1970; Marts and Wilkins, 1970) indicated that it should have greater potential than the adenosine triphosphate (ATP) bioluminescence method (Chappelle and Levin, 1968; McElroy et al., 1969) for monitoring bacteria levels in field applications where adequate laboratory facilities are not available.

The principle of the luminol chemiluminescence method for detecting bacteria is based on microbial activation of the oxidation of the luminol monoanion by hydrogen peroxide. The general reaction mechanism is shown in figure 1. In an aqueous alkaline solution, luminol (3-aminophthalhydrazide) is reported (Hodgson and Fridovich, 1973) to be in the form of the monoanion, which can be rapidly and energetically oxidized by hydrogen peroxide in the presence of the iron porphyrins contained in microorganisms (Erley et al., 1962; White and Roswell, 1972; Nikokavouras and Vassilopoulos, 1971). The intermediate products of oxidation could include free radicals such as hydroxyl, luminol, and O_2^- as well as luminol endoperoxide (Hodgson and Fridovich, 1973; Erley et al., 1962; White and Roswell, 1972; Nikokavouras and Vassilopoulos, 1971; Lee and Seliger, 1965). A final reaction product, the aminophthalate dianion, is formed in an electronically excited state which decomposes rapidly to the ground electronic state with loss of the excess energy as blue (430 nm) light. When the reaction takes place in the presence of excess reagents, the intensity of the light has been shown to be proportional to the concentration of bacterial porphyrins.

The choice of the chemiluminescence over the bioluminescence method was based on the advantages listed below, most of which are concerned with the greater adaptability of the chemiluminescence method to field applications. For example, the reagents are inexpensive, stable chemicals which do not require refrigeration or deep-freeze storage. Also, the preparation of the

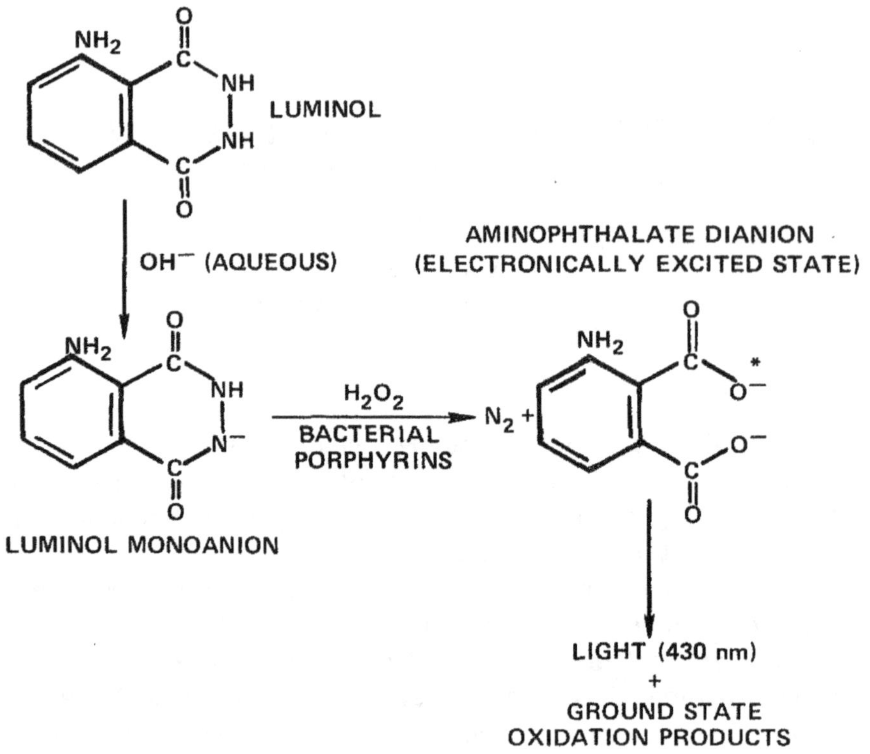

Figure 1. Detection of bacteria via chemiluminescence of luminol.

reagent and the analytical technique do not require the degree of care and manipulation that is needed in the bioluminescence method. Since the reagent is an alkaline solution which lyses the cells as rapidly as they are introduced, the separate lysing and extraction step is not required. Elimination of the prior lysing step, previously used in the chemiluminescence technique, was shown in this study to improve considerably the reproducibility and accuracy of the method in addition to simplifying it. When the cells were lysed with NaOH prior to analysis, very accurate timing between lysing and measurement was required for good reproducibility because the released porphyrins deteriorate rapidly in the presence of the lysed cells. In contrast, direct injection of the sample into the reagent eliminates the need for accurate timing.

The Aminco Chem-Glow Photometer (4-7441A), shown in figure 2, was used in all of these studies. It is an inexpensive, portable photomultiplier detector which gives a meter output showing the maximum light intensity produced when the sample is added to the reagent. In practice, one milliliter of the mixed reagent is introduced into the glass vial which is placed under the injection port opposite the photomultiplier tube in the light-tight rotatable reaction chamber. A volume of sample between 10 and 200 μl is injected into the reagent with the use of a mechanical injector attached to the syringe to ensure reproducible speed and sufficient force of injection for adequate mixing. The reproducibility is generally better than ± 5 percent.

96

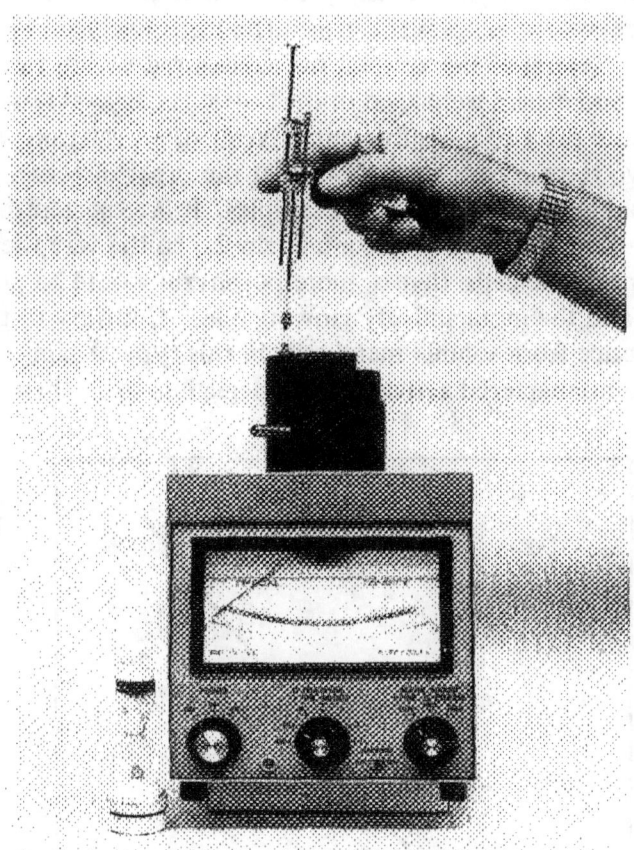

Figure 2. Aminco Chem-Glow Photometer, Shandon
Mechanical injector, and filtering equipment.

Since activation of the chemiluminescence reaction of luminol is not limited
to bacterial porphyrins, but can be promoted by metallic salts and some
organic compounds, the reaction, as such, is not specific for bacteria. However, by separating the bacteria from the nonbacterial portion of the sample
and by measuring and correcting for the light produced by the latter, the
method used is specific for bacteria. This is accomplished simply with the
use of a plastic syringe and bacterial membrane holder such as that shown to
the left of the instrument in figure 2. A few milliliters of the sample are manually pushed through the bacterial membrane having a pore size of 0.20 to
0.45 μm. A volume of bacteria-free filtrate, equal to that used for the total
sample, is injected into a fresh portion of reagent. The chemiluminescence
intensity produced by the filtrate is then subtracted from that given by the
total sample to yield the intensity due to the bacteria alone.

Results of a comparative study of the chemiluminescence and the plate count techniques on samples of cooling tower water are shown in figure 3. The six points represent samples taken at different times from two different cooling towers. Several of the samples were measured within two days after sampling, and others were aged in the bottle for several months at room temperature prior to measurement. Excellent correlation was found between the two tests in spite of the fact that the types of cells and stages of growth probably differed among the samples. It is expected that these factors would affect the porphyrin content similar to the way they affect the ATP content of the cells, that is, more porphyrin would be present in the larger cells and the more actively growing ones. Contrary to the expected interference from soluble metal salts in this type of sample, the emission due to nonbacterial activation was negligible in all these samples.

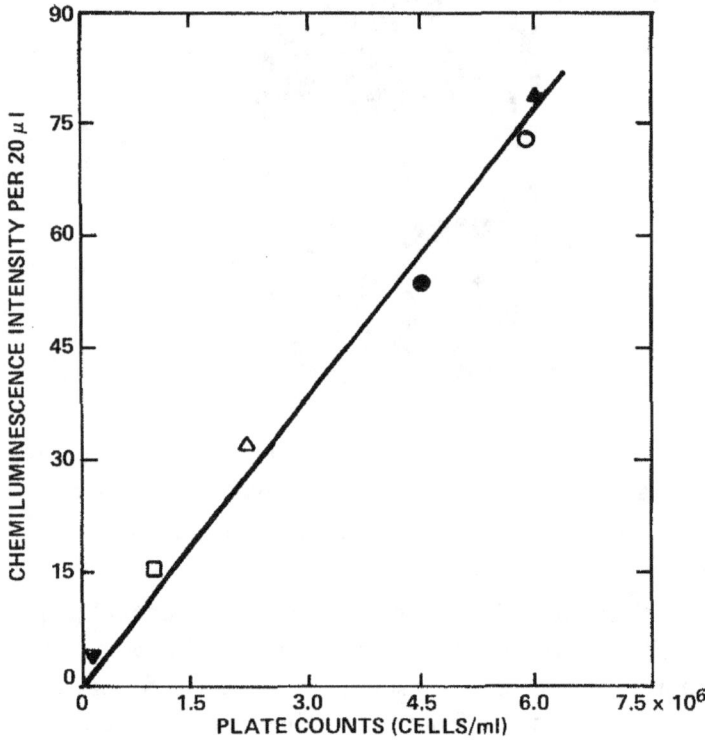

Figure 3. Cooling tower (CT) water.

Figure 4 illustrates the use of the chemiluminescence method in monitoring the effect of chlorine treatment in one of the cooling towers. Chlorine treatment was started at 9:30 a.m. and discontinued at about 3 p.m. Samples were collected every hour during treatment and every three hours after treatment was stopped. The decrease in cell population resulting from the treatment and the subsequent rise, beginning several hours after the treatment was stopped, is shown similarly by the plate counts and chemiluminescence measurements. In the initial study, an arbitrary correction was made for the chemiluminescence intensity due to the hypochlorite (ClO⁻)

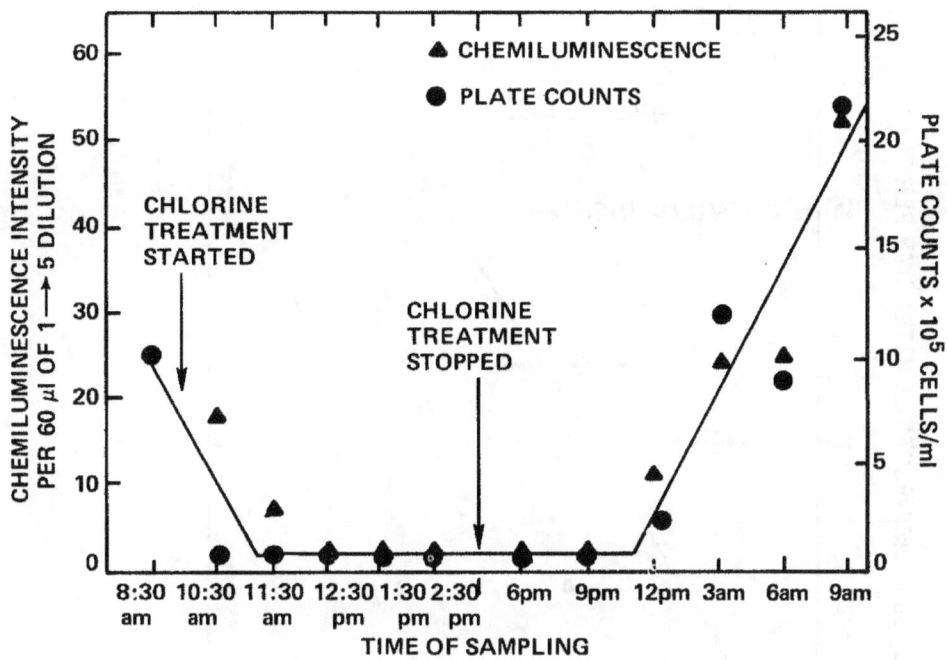

Figure 4. Chlorine treatment of cooling tower water.

present because the filtrate technique could not correct for it satisfactorily. In later studies, it was shown that interference by ClO⁻ could be eliminated completely by the addition of sodium thiosulfate to the sample.

The studies of cooling tower water show that the luminol chemiluminescence technique can be used to monitor changes in viable cell population both under normal conditions and during chlorine treatment. The limit of detection of bacteria by direct analysis, that is, without concentration of the cells, is estimated to be about 50,000 cells per milliliter for these samples based on plate-count analysis of the cell populations.

Good correlation between chemiluminescence and plate counts was also obtained in the analysis of process water used in paper mills. The results are shown in figure 5. In these samples, a considerable amount of paper pulp is present in the water. Since the microbial cells appear to be closely associated with the pulp, correlation between chemiluminescence and plate counts is dependent on identical sampling for both techniques so that the amount of pulp is the same. The range of replicates is shown by the horizontal bars for the plate counts and by the vertical bars for the chemiluminescence. Reproducibility was found to be considerably better for the latter. The limit of detection by the direct chemiluminescence method is estimated to be about an order of magnitude more for these samples than for the cooling tower, that is, about 500,000 cells per milliliter.

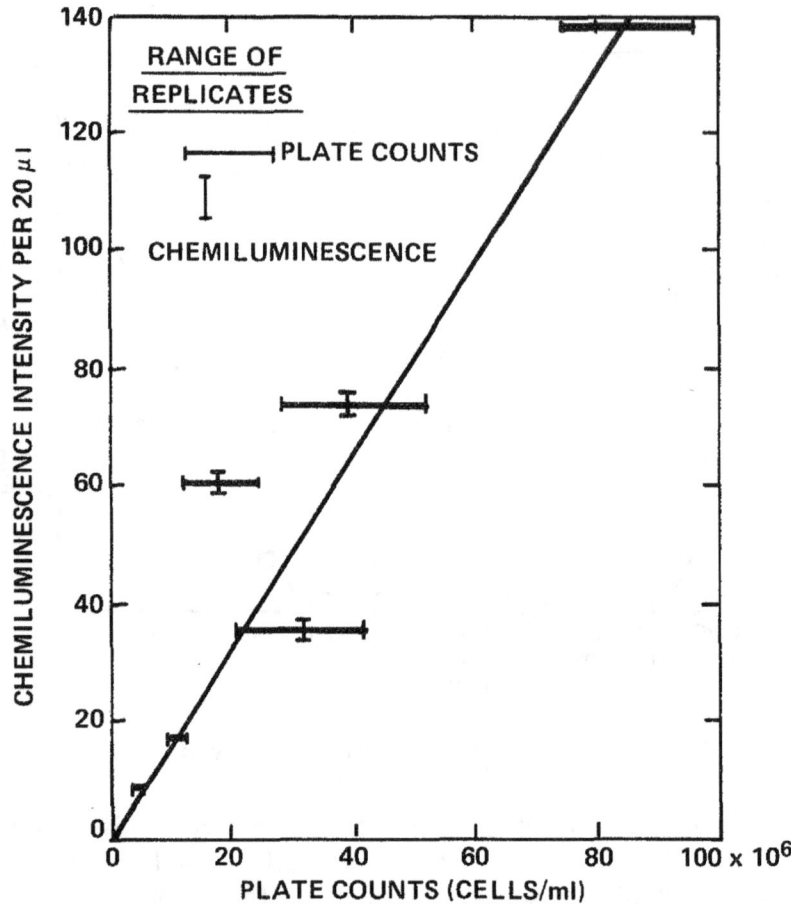

Figure 5. Process water containing paper pulp.

The chemiluminescence method has also shown good potential for monitoring the viable bacteria populations in activated sludge used in waste treatment plants to digest organic matter. Figure 6 shows the results of comparative tests on samples from the basin effluent of an industrial waste treatment plant taken over a period of several months. The amount of mixed liquor volatile suspended solids (MLVSS) is assumed to be indicative of the viability of the sludge. However, the fraction of the total MLVSS due to living microorganisms is variable, and it has been reported (McKinney, 1962; Patterson, 1970) that in some waste treatment plants it can often be as low as 25 percent. Considering the nonspecificity of the MLVSS test, correlation with the chemiluminescence intensities was better than expected. Considerably better correlation should be obtained when comparison is made with oxygen uptake, a test which gives a measure of the metabolic activity of the cells.

In addition to application to cooling tower water, paper mill water, and activated sludge, the method also has potential for applications such as:

100

- Bacterial cultures,
- Fermentation processes,
- Antibiotic efficacies,
- Process water, and
- Airborne bacteria.

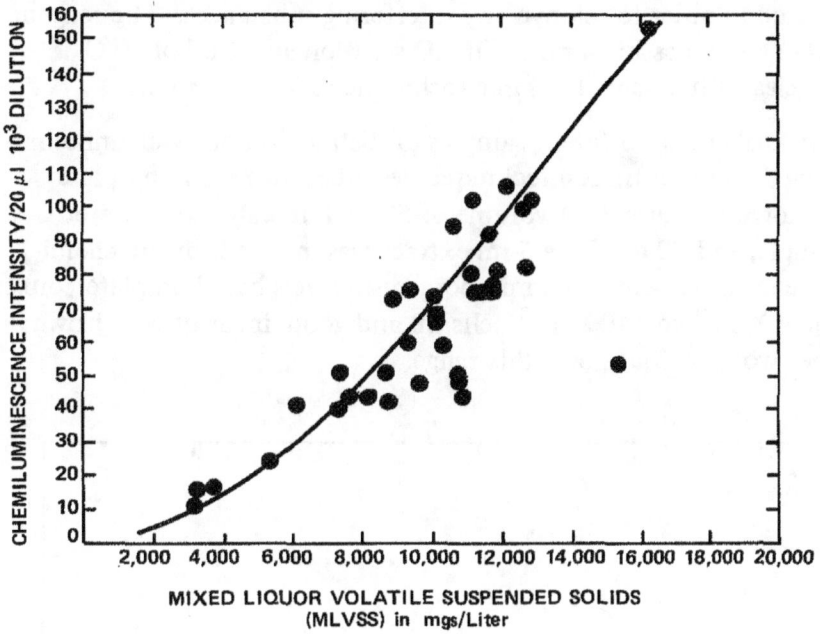

Figure 6. Chemiluminescence analysis of bacteria versus MLVSS
of activated sludge.

These include standardization of bacterial cultures, monitoring of fermentation processes, determination of the efficacy of antibiotic levels, testing for purity of process water, and determination of levels of airborne bacteria. In the latter two applications, the bacteria levels are expected to be below that which can be measured by direct analysis. Therefore, a concentration technique is required. It has been shown that by collecting the cells contained in a large volume of water on a bacterial filter and extracting the active components with a 90:10 DMSO:H_2O solution, lower levels of bacteria can be measured.

In the technique used for low levels of bacteria, a volume of water ranging from 50 milliliters to several liters, depending on the level of bacteria, is filtered through a 47-mm, 0.2-μm, or 0.45-μm Acropor AN Gelman membrane by suction using a Sartorius Membranfilter plastic filtration system or equivalent apparatus. The membrane is then transferred to a 47-mm SWINNEX® Millipore filter holder. Exactly 2 ml of the DMSO (dimethyl-sulfoxide) solution are added directly with a pipet through the top opening of the filter holder. After several minutes of contact with the filter, the DMSO solution is manually forced through the membrane by means of a plastic syringe. The procedure is repeated with a second 2-ml portion of DMSO solution and finally with 1 ml of sterile water. The total filtrate is then adequately mixed, and an aliquot is injected directly into the reagent. Sterilization of the membrane and filter system is not required. However, at least 5 ml of sterile water should be filtered through the membrane before it is used in order to remove any interfering substances. The reagent blank is obtained by passing 4 ml of DMSO solution and 1 ml of H_2O through the clean filter using the same technique as for the sample.

Comparative analyses of different samples of distilled water by chemiluminescence using the concentration technique described above and by plate counts are shown in figure 7. A volume of 500 ml of water was filtered for each sample, and 50 μl of the 5-ml extract was injected for the chemiluminescence measurement. The microorganism levels based on plate counts ranged from 6×10^3 to 140×10^3 cells/ml and good linearity was shown between the two techniques over this range.

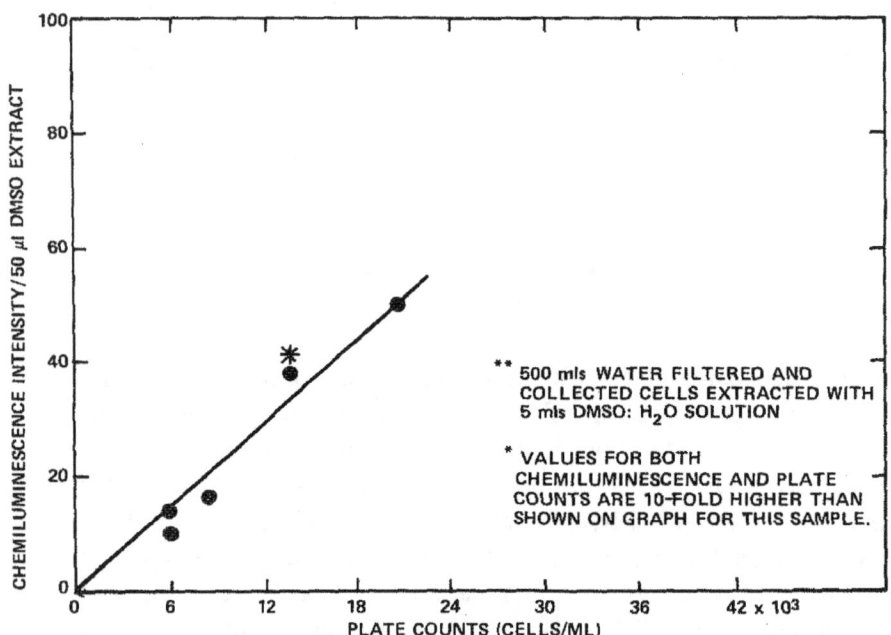

Figure 7. Distilled water: chemiluminescence** versus plate counts.

Testing for airborne bacteria could be accomplished by an impinger-type collection in water followed by concentration and extraction as described for distilled water. The lowest limit of detection of the cells in distilled water was 100 cells/ml, measured by concentrating the cells present in several liters of water. Modification of the method for application to lower

102

levels of bacteria by increasing the sensitivity of the reagent and improving the extraction procedure is being investigated.

REFERENCES

Chappelle, E. W. and G. V. Levin, "The Use of the Firefly Bioluminescent Assay for the Rapid Detection and Counting of Bacteria," *Biochem. Med.*, 2, 1968, p. 41-52.

Erley, L., W. F. Pickering, and C. L. Wilson, "Chemiluminescence of Luminol and Haemin," *Talanta*, 9, 1962, p. 653-659.

Hodgson, E. K. and I. Fridovich, "The Role of O_2^- in the Chemiluminescence of Luminol," *Photochem. and Photobiol.*, 18(6), 1973, p. 451-455.

Lee, J. and H. H. Seliger, "Absolute Spectral Sensitivity of Phototubes and the Application to the Measurement of the Absolute Quantum Yields of Chemiluminescence and Bioluminescence," *Photochem. and Photobiol.*, 4, 1965, p. 1015-1048.

Marts, E. C. and J. R. Wilkins, "Simple Chamber Facilitates Chemiluminescent Detection of Bacteria," Tech. Brief B-70-10525, NASA Langley Research Center, Hampton, Virginia, 1970.

McElroy, W. D., H. H. Seliger, and E. H. White, "Mechanism of Bioluminescence, Chemiluminescence, and Enzyme Function in the Oxidation of Firefly Luciferin," *Photochem. and Photobiol.*, 10, 1969, p. 153-170.

McKinney, R. E., *Microbiology for Sanitary Engineers*, McGraw Hill Book Company, Incorporated, 1962, p. 167-169.

Nikokavouras, J. and G. Vassilopoulos, *Z. fur Physik Chemie Neue Folge*, 75(3/4), 1971, p. 180-184.

Oleniacz, W. S., M. A. Pisano, M. H. Rosenfeld, and R. F. Elgart, "Chemiluminescent Method for Detecting Microorganisms in Water," *Environ. Sci. and Tech.*, 2(11), 1970, p. 1030-1033.

Patterson, J. W., P. L. Brezonik, and H. D. Putnam, "Measurement and Significance of Adenosine Triphosphate in Activated Sludge," *Environ. Sci. and Tech.*, 4, 1970, p. 569-575.

White, E. H. and D. F. Roswell, "The Chemiluminescence of Organic Hydrazides," *Res. Accounts*, 3, 1972, p. 54-62.

SIMPLE PHOTOMETER CIRCUITS USING MODULAR ELECTRONIC COMPONENTS

John E. Wampler
Bioluminescence Laboratory
Department of Biochemistry
University of Georgia
Athens, Georgia

INTRODUCTION

During the last several years we have seen a tremendous increase in the variety and performance of solid-state electronics. Many modular components have become available at attractive prices, which give the photobiologist workable alternatives to commercial light measurement instrumentation. More importantly, the availability and flexibility of these modules allows us to inexpensively construct specialized measurement devices.

The basic components of light measurements are a detector, a high voltage power supply (if the detector is a photomultiplier), an amplifier, and a read-out system. The detector can be a photodiode, phototransistor, photocell, pyrometer, or photomultiplier tube. For low light level measurements the photomultiplier is the detector of choice, but it requires a high voltage supply. No longer must the high voltage supply be a large piece of equipment; in fact a small potted module can now serve this function for most photomultipliers, and the power requirement of the module itself is low voltage d.c. The amplifier can also be a small modular component, as can the read-out system. In fact, the variety of amplifiers, digital meters, and counters as well as other more specialized modules (for example, frequency converters, multiplier-divider modules, log modules) allows us to custom design instruments for many specialized applications. This paper describes an instrument for general bioluminescence assay use with special capabilities for an assay requiring the measurement of the ratio of two light signals. The individual circuits, as well as the considerations of overall design and construction, are discussed.

OPERATIONAL AMPLIFIERS

The key component in design of the current amplifier for a photomultiplier, and in many circuits which are useful for read-out applications, is the operational amplifier (op-amp) (Graeme et al., 1971; Philbrick/Nexus Research, 1969; Vassos and Ewing, 1972). Op-amps are differential linear d.c. amplifiers with very high gain which can be utilized to implement a wide variety of analog signal processing functions. There are many types of modular amplifiers now available: for high input impedance applications, FET input amplifiers, and for low drift application, chopper stabilized op-amps. In addition, for special requirements for high current or voltage capabilities or high frequency response selected units are available. Figure 1 illustrates how some simple arrangements of components connecting the two inputs and the output of an op-amp can perform useful functions.

Referring to figure 1A, all of these circuits consist of a single-ended input and feedback from the amplifier output to one of its inputs. The overall performance of the circuit is dictated by this feedback and not the nature of the amplifier module itself. The simplest circuit shown (figure 1B) is the so-called voltage follower amplifier; in this application a voltage input signal is simply buffered by the amp. Many op-amps have high input impedance (particularly FET input op-amps), thus they can be used to buffer the output of a voltage source with a low current capability and drive a read-out device which requires a relatively high current. In circuit configurations of the type of figures 1C and 1D, the inverting input of the amplifier is used. This input is held by the amplifier at essentially ground potential by feedback; that is, the sum of currents flowing into the inverting input is zero. The current from the signal source is equalled by the flow of current from the output through the feedback components. The inverting input is often called the summing junction because the amp acts to sum the currents at this point to zero when appropriate feedback is applied. In the circuit of figure 1C, an output voltage is produced which is proportional to the input current; the proportionality constant is simply the feedback resistance times minus one (since the feedback current must be of equal magnitude but opposite in sign to the signal current). When a signal is from a voltage source rather than a current source, it is a simple matter to convert the voltage to a current for input. This is accomplished by connecting the voltage source to the summing junction through a resistor as in figure 1D. The ratio of the output voltage to that of the signal source voltage is now proportional to the ratio of the two resistors as shown. In each of these circuits a capacitor has been introduced across the feedback resistor to limit the frequence response. The time constant in seconds, which describes the response of the complete circuit, is simply the product of R_1 and C. A useful amplifier for the output of a photomultiplier employs aspects of both 1C and 1D (figure 2). Here the negative current from a photomultiplier anode is transduced by the op-amp into a voltage. With R_1 equal to 10 MΩ (10^7 ohm), -10^{-8} amps would give a 0.1-V output signal. In order to cancel any dark current from the photomultiplier, a positive voltage can be selected using R_3 to feed a positive offsetting current into the summing junction through R_2.

PEAK HOLDING AMPLIFIER

A very useful circuit for bioluminescence assays is a peak holding circuit (Graeme et al., 1971; Philbrick/Nexus Research, 1969) (figure 3). Since initial

106

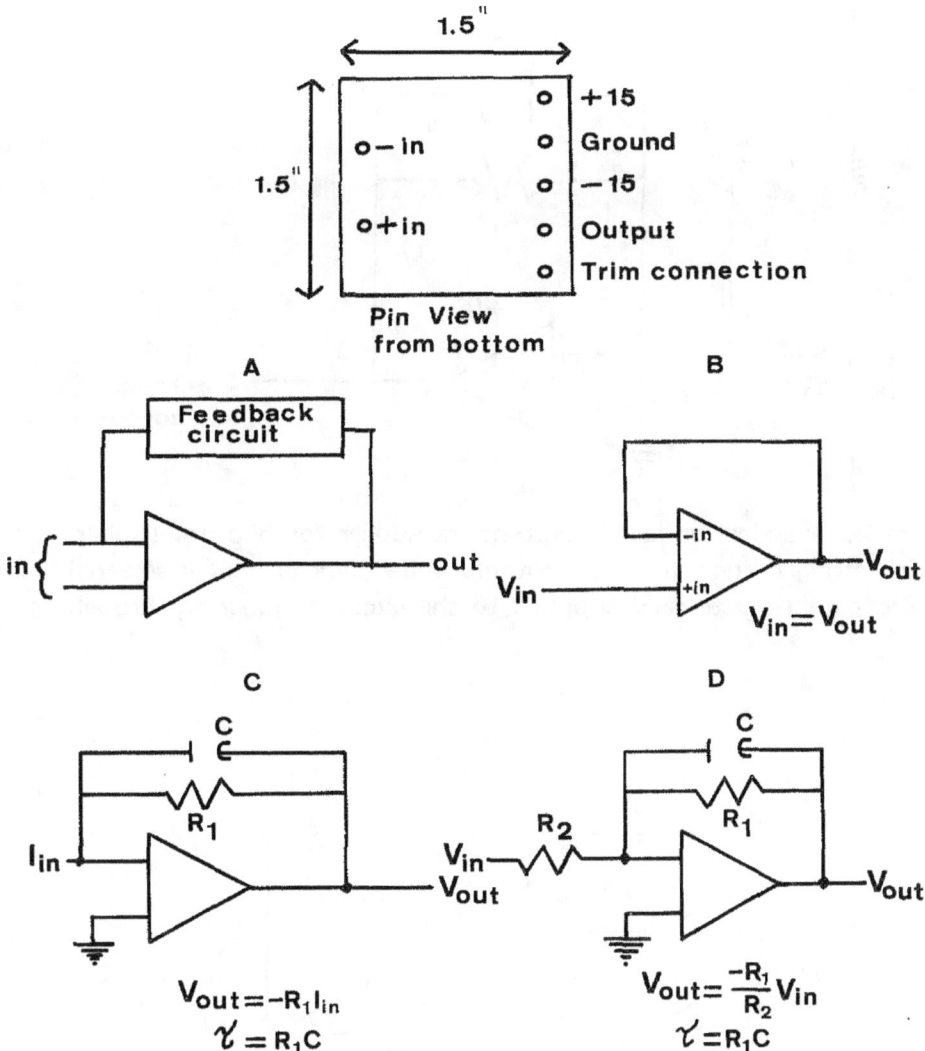

Figure 1. Operational amplifier circuits. A typical potted op-amp contained in a small module would have seven connecting pins as shown. Some simple circuits are (A) a general circuit, (B) a voltage follower, (C) a current-to-voltage transducer, and (D) an inverting voltage amplifier.

rate in many assays can be equated to the height of the flash peak, the peak hold allows the experimenter to read the peak value at his leisure. This simple circuit employs two op-amps in a noninverting configuration. The rate of change of the output after the peak is passed is determined by how leaky capacitor C_1 is and by the input impedance of A2. For this reason an FET op-amp is used for A2. Other peak hold circuits are discussed in the references given.

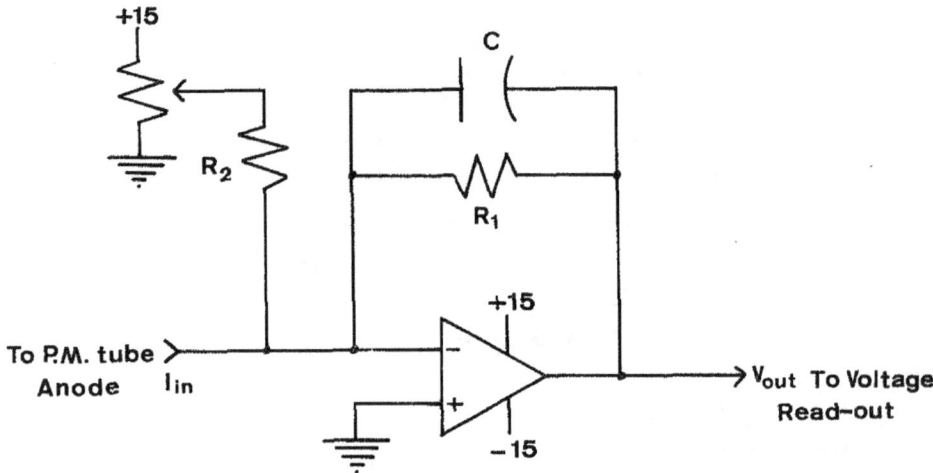

Figure 2. Practical voltage-to-current transducer for photomultiplier. An offsetting current for the photomultiplier dark current is selected via the potentiometer and supplied to the summing junction through R_2.

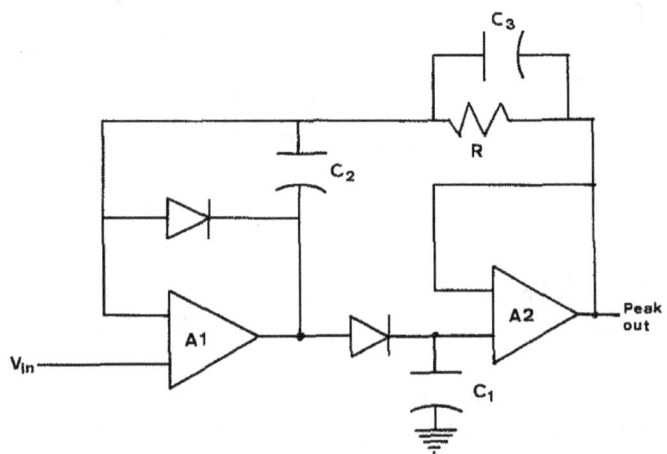

Figure 3. A noninverting peak hold circuit from Graeme et al. (1971). Two amplifiers are used in non-inverting configurations.

INTEGRATION

A second useful function for bioluminescence assays is signal integration. While analog methods can give a good integration on short time scales (Philbrick/Nexus Research, 1969), digital methods are best for long term integration. A convenient and easy method of integration is to convert a signal voltage into a pulse train and then simply count the pulses. Since

108

linear voltage-to-frequence converters are now available in modules similar to op-amps, this circuitry is easily implemented (figure 4).

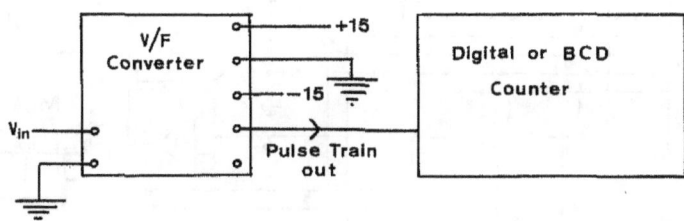

Figure 4. Using a modular voltage-to-frequency converter to drive a counter for signal integration.

POWER SUPPLIES

Most operational amplifiers require ±15-V power supplies. These are now available in a variety of modules. Many of the op-amp manufacturers also offer power supplies. Digital logic requires +5 V, also available in modules.

The only remaining power requirement is that of the photomultiplier. Several manufacturers make small modular d.c.-to-d.c. converters which will convert low voltage d.c. into the high voltage d.c. necessary for the photomultiplier. RCA has recently introduced integrated photomultiplier modules (C350001C/PF1004C Integrated Photodetection Assemblies) containing the photomultiplier, dynode chain, and power supply integrated into a single package. Generally, d.c. converters can be powered by the +15-V power supply used for the op-amps.

A GENERAL PHOTOMETER CIRCUIT WITH RATIO CAPABILITY

The circuit in figure 5 was designed for both general bioluminescence assays and for a specialized assay involving detection of the ratio of blue light to green light in coelenterate in vitro assays (Wampler et al., 1971; 1973). Here we use an autoranging digital voltmeter (Datel DM2000AR*) as a read-out device, a voltage-to-frequency converter-integrator circuit, and a multiplier-divider module to obtain the ratio. Table 1 lists the parts for this instrument. Signals 1 and 2 for the ratio assay are obtained from a detector module (figure 6). For regular assays with low light levels, an end window tube (EMI 9635B) is used. In this case the power supply is a d.c.-to-d.c. converter (PMSI-3A†). The housing for this detector (figure 7) is machined from

---

* Datel Systems, Inc., 1020 Turnpike Street, Canton, Massachusetts.
† Del Electronics, Mt. Vernon, New York.

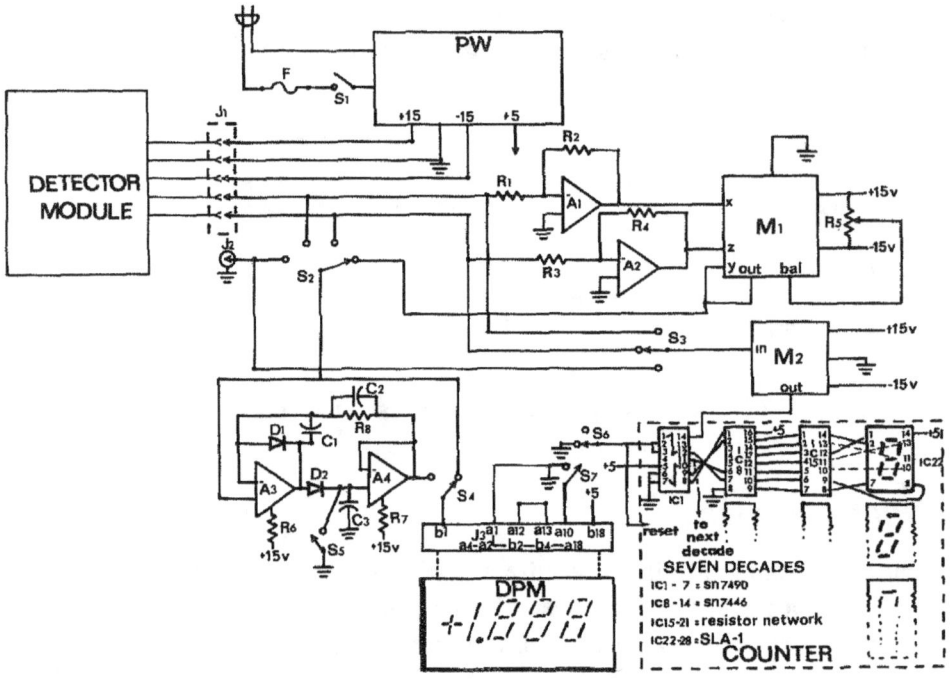

Figure 5. Circuit diagram of photometer. The parts and the values of the components are in table 1.

aluminum with a planar shutter to allow close proximity between the tube face and the sample.*

The signals from the photomultiplier tubes are transduced by the circuit of figure 7, and the output signal of the op-amp is transmitted to the read-out instrument via a multiconductor cable which also acts to carry the ±15-V power to the detector. Switch S_2 allows the digital voltmeter (DVM) to measure the signal from either photomultiplier or the ratio from the 426A module, and S3 allows integration of either signal 1 or signal 2. With S4, the peak hold function can be selected for display on the meter instead of intensity. S5 resets the peak circuit while S6 resets the integrator. The DVM and the integrator can also examine an input signal from other voltage sources using the front panel BNC connector marked TEST.

For the ratio measurement, using the 426A module, the denominator must be larger than or equal to the numerator and both must be negative for a positive output; the output signal read by the DVM is 10 X signal 1/signal 2. Because signals 1 and 2 are positive voltages, they must be inverted. The inverting amplifiers, A1 and A2, are integrated circuit op-amps with precision resistors for unity gain. For the best measurement, the high voltage of tube 2 should be adjusted until signal 2 is near 10 V, then signal 1 adjusted to an appropriate value.

---

* Faini, G. L., unpublished, 1975.

110

Table 1

Parts List

Symbol	Item and Source or Description	Approximate Cost
Read-out Chassis:		
A1 and A2	741 Integrated circuit op-amps	$ 2.00/ea
A3 and A4	Analog Devices 40J—FET input general-purpose op-amps: Analog Devices, Inc., Cambridge, Mass.	$ 12.00/ea
C1 and C2	0.1 μF capacitors	---
C3	0.47 μF capacitor—selected low leakage capacitor	$ 5.00
D1 and D2	diodes	---
DPM	Autoranging digital panel meter: Datel Systems, Inc., Canton, Mass.	$160.00
F	3 amp fuse	---
IC_1–IC_7	SN4790 decade counter	$ 1.00/ea
IC_8–IC_{14}	SN7446 seven segment driver	$ 1.00/ea
IC_{15}–IC_{21}	Allan Brady 7443 or similar 150 Ω resistor array	$ 2.00/ea
IC_{22}–IC_{28}	OPCOA SLA-1 LED display: OPCOA, Edison, New Jersey	$ 2.00/ea
J_1	5 pin military connector	$. 5.00
J_2	BNC connector	---
J_3	P.C. board connector supplied with DPM	---
M1	Multiplier module 426A: Analog Devices Inc.	$ 45.00
M2	Voltage-to-frequency converter 4701: Teledyne Philbrik, Dedham, Mass.	$ 60.00
PW	Model 2R-70T triple power supply 115: +5 V d.c. power tec, Chatsworth, Calif.	$110.00
R1-R4	10K, 0.1%, ¼ W resistors	---
R5	20K trim pot	---
R6 & R7	500 Ω, 10%, ¼ W resistors	---
R8	100 K, 1%, ¼ W resistors	---
S1	5 amp, lighted push button switch	$ 3.00
S2	4 position rotary switch	---
S3	3 position rotary switch	---
S4	SPDT toggle switch	---
S5 & S7	Normally open momentary push button switch	---
S6	Normally closed momentary push button switch	---
Miscellaneous:	Wire, cables, sockets, chassis, power cord, and so forth	
Detector Module 1:		
PM_1 & PM_2	RCA integrated photodetector assemblies	$120.00/ea
A_5 & A_6	Chopper stabilized op-amps model 233J: Analog Devices, Inc.	$ 60.00/ea

Table 1 (continued)

Symbol	Item and Source or Description	Approximate Cost
Detector Module 1 (continued):		
R_9 - R_{12}	10K, 10-turn potentiometers	$ 6.00/ea
R_{13} - R_{14}	10 MΩ, 1% resistors	---
R_{15} - R_{16}	10 MΩ resistors	---
C_4 & C_5	0.01 μF capacitor	---
B	Beam splitter: Edmund Scientific Co., Barrington, N.J.	$6.00 - $20.00
Detector Module 2:		
PM_3	EMI 9635 B photomultiplier: Gencom Division, Plain View, N. Y.	$205.00
PW_2	High voltage power supply—d.c.-to-d.c. converter—PMSI-3A: Del Electronics, Mt. Vernon, N. Y.	$ 90.00
A_7	Low bias FET op-amp Model 1029: Teledyne Philbrick, Dedham, Mass.	$ 45.00
R_{17}	1000 ohm, 10-turn potentiometer	$ 6.00
R_{18}	10 MΩ resistor	---
R_{19}	10 K, 10-turn	$ 6.00
R_{20}	50 K trim potentiometer	---
R_{21}	100 MΩ, 1% resistor	---
C_6	0.001 μF capacitor	---

All calibration adjustments can be performed from the front panel using the TEST input. The bezel of the meter conceals its internal offset, gain, and zero adjustments. No adjustments are required of the integrator circuit or the peak hold.

The Datel DVM is autoranging, that is, it shifts between read-out ranges of 199.9 mV, 1.999 V, and 19.99 V full scale as the signal input dictates. The meter's hold function, S7, allows leisurely reading of a value, and with three and one-half significant figures and three scales, the range of light measurements for a particular high voltage setting is greater than 10^4 for a reading of three significant figures.

The time constant on the detector current transducers (figure 7) is 0.1 s and a chopper stabilized op-amp was chosen to minimize long term drift. Changing the feedback resistor R would also change the gain, and switchable values could be used to increase the dynamic range of the instrument for a fixed voltage setting.

The power supply chosen, while not as compact as some available, has considerable current output capabilities and is cheaper than less powerful, but smaller, potted modules.

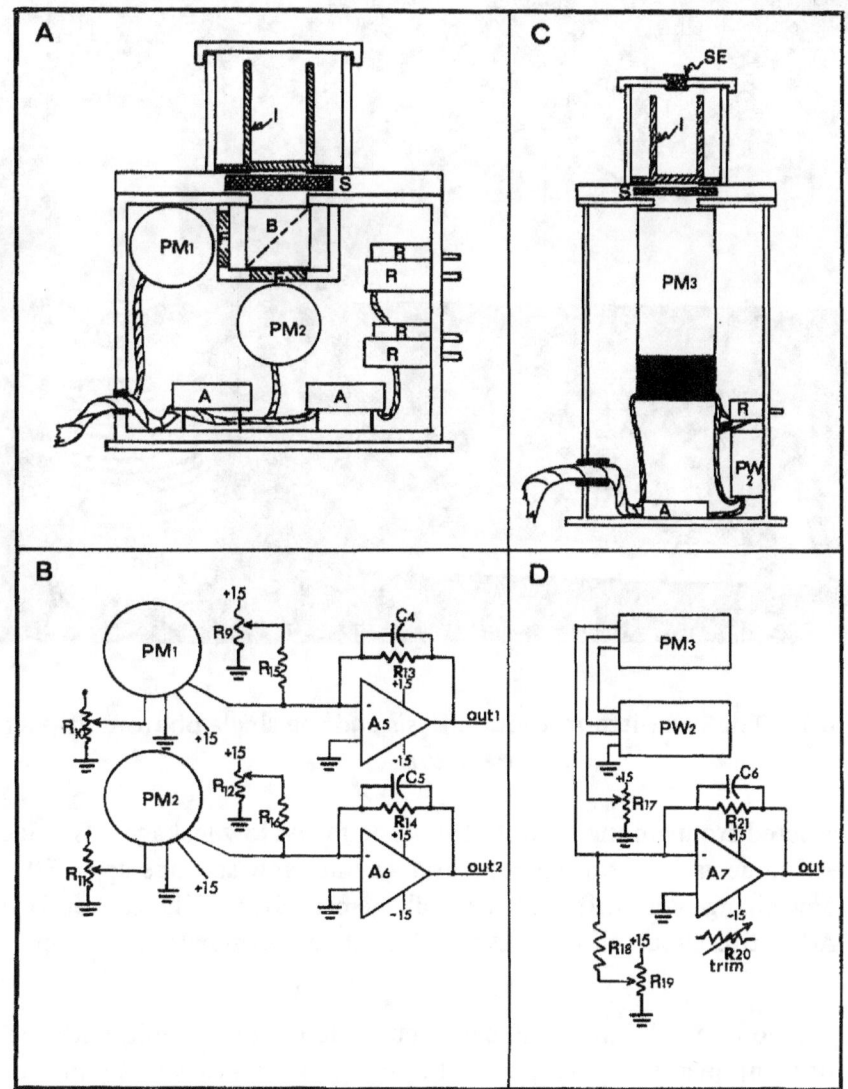

Figure 6. Diagram of the detector assemblies. A. The two-photomultiplier assembly used to assay the ratio of blue light to green light in coelenterate bioluminescence. In this instrument, B is a beam splitter, S the planar shutter mechanism, and the filters F are interference filters (470 nm for the blue component and 510 nm for the green one). B. The circuit of Assembly A. C. The single photomultiplier assembly. D. The circuit diagram of Assembly C. The electronic parts used in these instruments are detailed in table 1. I is a lucite insert used to hold specific assay vials, SE a syringe septum, and S the planar shutter.

The choice of component values in the peak hold circuit allows tracking of a full scale change (0 to 10 V) with a rise time of less than 10 ms. Capacitor

113

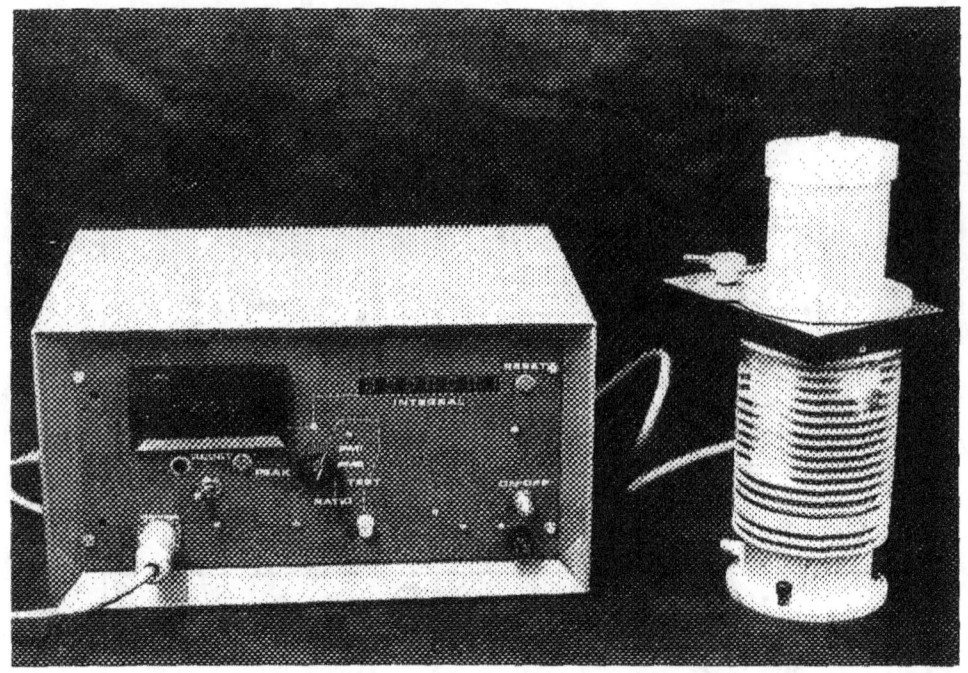

Figure 7. The complete read-out chassis and the single photomultiplier detector.

C_3 was selected from among several of its type for its low leakage. The decay of the peak value varied from 10 percent decay in a few seconds up to 30 min for the same change with different low-leakage capacitors. The socket for the op-amp A4 of this circuit should have well insulated terminals to prevent leakage.

While a printed circuit board could easily be made for this circuit, with the number of components involved, point-to-point wiring is easy and convenient, particularly if all amps and modules are mounted in sockets. Soldered connections were used except in wiring the counter. For the many interconnections of the counter and to facilitate packing, it was wired by wirewrap technique using wirewrap pin sockets and 30-gage Teflon insulated wire.

VARIATIONS OF THE BASIC DESIGN

For some measurements certain changes in the circuit of figure 5 might be made. A timed integration might be useful for some assays. A simple digital timer (Collins, 1974) could be arranged to gate the integrator on and off at preset times. Another variation would be to incorporate the voltage-to-frequency converter into the detector housing and then transmit the light intensity information as a pulse train. This would allow long distance transmission of the luminescence information in field applications where the detector cannot be near the read-out device, such as in a probe for underwater luminescence.

114

ACKNOWLEDGMENT

The development of this photometer and its predecessors has been greatly aided by interaction with several of my colleagues. I am particularly grateful to Drs. Cormier, Lee, Faini, and DeSa of this department for their useful advice. This work was supported in part by NSF grant GB-43804.

REFERENCES

Collins, J. D., *Popular Electronics,* April 1974, p. 47.

Graeme, J. G., G. E. Tobey, and L. P. Huelsman, eds., *Operational Amplifiers Design and Applications*, McGraw Hill Book Company, 1971.

Philbrick/Nexus Research, Applications Manual for Operational Amplifiers, Philbrick/Nexus Research, Dedham, Mass., 1969.

Vassos, B. H. and G. W. Ewing, *Analog and Digital Electronics for Scientists*, Wiley-Interscience, 1972.

Wampler, J. E., K. Hori, J. W. Lee, and M. J. Cormier, *Biochem.*, **10**, 1971, p. 2903.

Wampler, J. E., Y. D. Karkhanis, J. G. Morin, and M.J. Cormier, *Biochim. Biophys. Acta*, **314**, 1973, p. 104.

"The aeronautical and space activities of the United States shall be conducted so as to contribute . . . to the expansion of human knowledge of phenomena in the atmosphere and space. The Administration shall provide for the widest practicable and appropriate dissemination of information concerning its activities and the results thereof."

—NATIONAL AERONAUTICS AND SPACE ACT OF 1958

NASA SCIENTIFIC AND TECHNICAL PUBLICATIONS

TECHNICAL REPORTS: Scientific and technical information considered important, complete, and a lasting contribution to existing knowledge.

TECHNICAL NOTES: Information less broad in scope but nevertheless of importance as a contribution to existing knowledge.

TECHNICAL MEMORANDUMS: Information receiving limited distribution because of preliminary data, security classification, or other reasons. Also includes conference proceedings with either limited or unlimited distribution.

CONTRACTOR REPORTS: Scientific and technical information generated under a NASA contract or grant and considered an important contribution to existing knowledge.

TECHNICAL TRANSLATIONS: Information published in a foreign language considered to merit NASA distribution in English.

SPECIAL PUBLICATIONS: Information derived from or of value to NASA activities. Publications include final reports of major projects, monographs, data compilations, handbooks, sourcebooks, and special bibliographies.

TECHNOLOGY UTILIZATION PUBLICATIONS: Information on technology used by NASA that may be of particular interest in commercial and other non-aerospace applications. Publications include Tech Briefs, Technology Utilization Reports and Technology Surveys.

Details on the availability of these publications may be obtained from:

SCIENTIFIC AND TECHNICAL INFORMATION OFFICE

NATIONAL AERONAUTICS AND SPACE ADMINISTRATION
Washington, D.C. 20546